Good Science—*That's Easy to Teach*

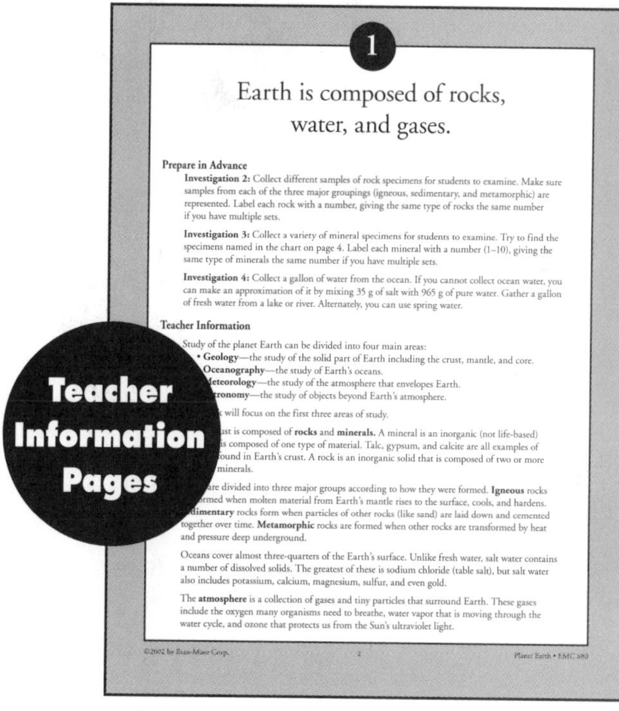

- the concept to be studied
- items to obtain or prepare in advance
- background information

- reproduce or make into a transparency

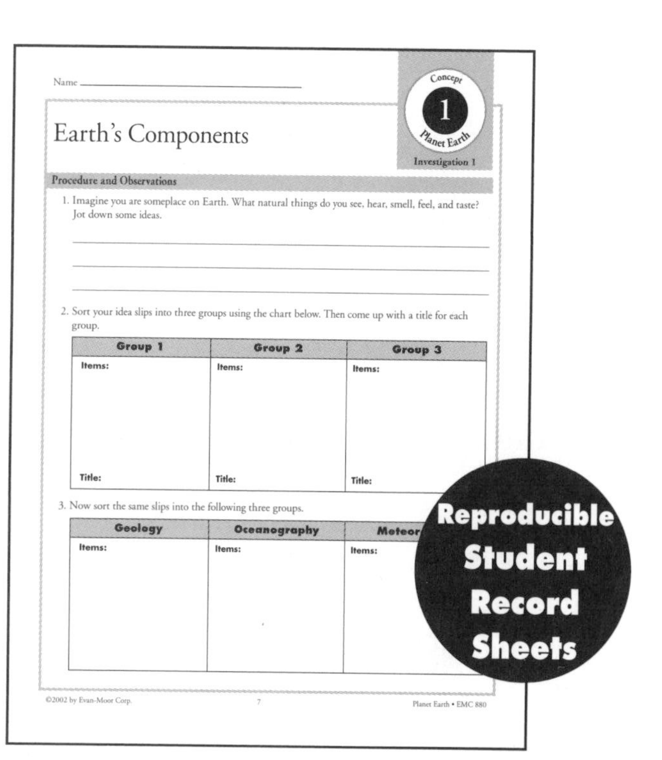

1

Earth is composed of rocks, water, and gases.

Prepare in Advance

Investigation 2: Collect different samples of rock specimens for students to examine. Make sure samples from each of the three major groupings (igneous, sedimentary, and metamorphic) are represented. Label each rock with a number, giving the same type of rocks the same number if you have multiple sets.

Investigation 3: Collect a variety of mineral specimens for students to examine. Try to find the specimens named in the chart on page 4. Label each mineral with a number (1–10), giving the same type of minerals the same number if you have multiple sets.

Investigation 4: Collect a gallon of water from the ocean. If you cannot collect ocean water, you can make an approximation of it by mixing about 1.2 oz. (35 g) of salt with about 4 cups (965 mL) of water. Gather a gallon of fresh water from a lake or river. Alternately, you can use spring water.

Teacher Information

Study of the planet Earth can be divided into four main areas:

- **Geology**—the study of the solid part of Earth including the crust, mantle, and core.
- **Oceanography**—the study of Earth's oceans.
- **Meteorology**—the study of the atmosphere that envelopes Earth.
- **Astronomy**—the study of objects beyond Earth's atmosphere.

This book will focus on the first three areas of study.

Earth's crust is composed of **rocks** and **minerals.** A mineral is an inorganic (not life-based) solid that is composed of one type of material. Talc, gypsum, and calcite are all examples of minerals found in Earth's crust. A rock is an inorganic solid that is composed of two or more different minerals.

Rocks are divided into three major groups according to how they were formed. **Igneous** rocks are formed when molten material from Earth's mantle rises to the surface, cools, and hardens. **Sedimentary** rocks form when particles of other rocks (like sand) are laid down and cemented together over time. **Metamorphic** rocks are formed when other rocks are transformed by heat and pressure deep underground.

Oceans cover almost three-quarters of the Earth's surface. Unlike fresh water, salt water contains a number of dissolved solids. The greatest of these is sodium chloride (table salt), but salt water also includes potassium, calcium, magnesium, sulfur, and even gold.

The **atmosphere** is a collection of gases and tiny particles that surround Earth. These gases include the oxygen many organisms need to breathe, water vapor that is moving through the water cycle, and ozone that protects us from the Sun's ultraviolet light.

Rock Types

Sedimentary Rock

Formed when particles of other rocks are deposited in layers and then compacted together over time.

Photo by Nick Ferris

Metamorphic Rock

Formed when existing rock is changed into another type of rock by extreme heat and/or pressure.

Photo by Nick Ferris

Igneous Rock

Formed when molten material from deep within the Earth rises to the surface, cools, and hardens.

Photo by Nick Ferris

Mineral Properties Chart

Mineral	Color	Density (g/cm³)	Luster	Hardness	Other
Calcite	colorless to white	2.7	glassy	3	bubbles with acid
Chlorite	green	3.0	glassy or pearly	2–2.5	flaky
Copper	copper red	8.9	metallic	2.5–3	conducts electricity
Galena	silver or gray	7.5	metallic	2.5	often in cube shape
Graphite	black or gray	2.3	sort of metallic	1–2	feels greasy
Gypsum	white, pink, gray, or colorless	2.3	glassy, silky, or pearly	1–2.5	
Halite	gray or colorless	2.2	glassy	2.5–3	salty taste
Magnetite	black	5.2	metallic	5–6	magnetic
Muscovite	gray, brown, or colorless	2.9	glassy or pearly	2–2.5	flaky
Olivine	olive green	3.2	glassy	6.5–7	
Pyrite	yellowy gold	5	metallic	6–6.5	turns black when crushed
Quartz	white or colorless	2.6	glassy or waxy	7	usually in crystals
Talc	white or gray	2.7	pearly or greasy	1–1.5	very soft

Layers of the Atmosphere

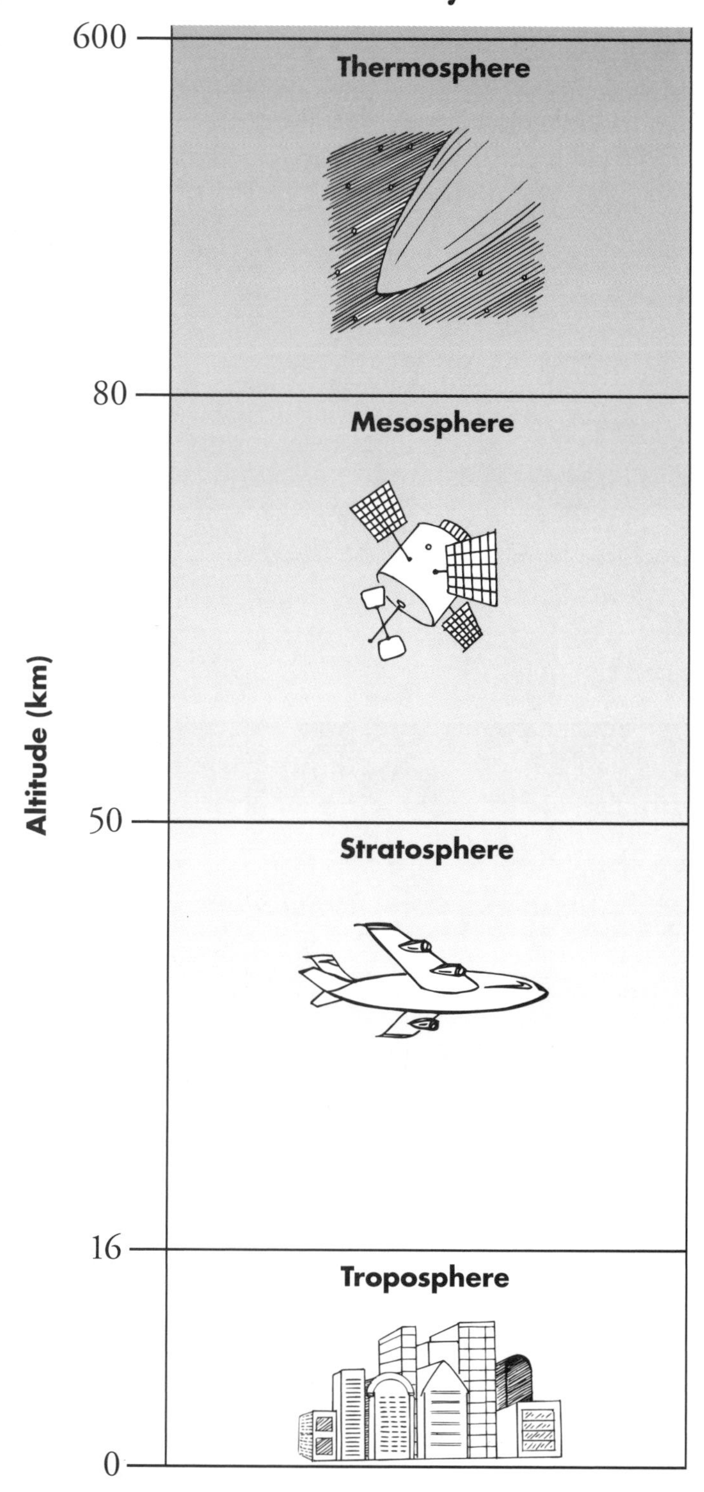

Gas particles in the upper part of the **thermosphere** absorb X rays and gamma rays from the Sun, making them electrically charged. Gases also absorb solar energy, raising the temperature in this layer.

The **mesosphere** is the coldest layer of the atmosphere.

Air in the **stratosphere** is very thin and dry. In the upper layers of the stratosphere, ozone absorbs energy from the Sun, warming the air particles.

Most of the atmosphere's gases and particles are found in the **troposphere.** All of Earth's weather occurs in this layer. Winds move heat and moisture around the globe.

Investigation 1

Earth's Components

Materials

- student record sheet on page 7, reproduced for each student
- small slips of paper
- three pieces of chart paper, one labeled "Geology," one labeled "Oceanography," and one labeled "Meteorology"
- tape

Steps to Follow

1. Ask students to imagine a favorite place they like to go. Have students close their eyes and picture every detail of where they are. Remind them to use all of their senses. What natural objects do they see; what natural sounds do they hear; what natural odors do they smell; what natural objects do they feel by touch; what natural flavors do they taste?
2. Have students record their ideas on their record sheets. Call on several individuals, one at a time, and have them share one or two of their responses.
3. Divide the class into groups of three. Distribute slips of paper to each group. Make sure each group has lots of slips.
4. Introduce the topic of "Things that make up the Earth." Turn the groups loose to brainstorm for 10 to 15 minutes. They should place individual ideas on one slip of paper (for example, "Rocks," "Trees," or "Water"). After an idea is written down, have students place the slip of paper in a pile in the middle of the table. Then have them move on to another idea and another slip of paper.
5. When students have run out of ideas, have each group sort their slips of paper into piles based on similarities. Stress that the fewer numbers of piles the better. Three is preferable. Have them list the items in each group on their record sheets.
6. Ask each group to come up with a descriptive title for each of their piles (one or two words). Have them record the titles on their record sheets.
7. Tape the labeled chart paper on the walls of your classroom.
8. Explain to students that one way scientists sort items on Earth is according to whether they are part of the Earth's rocky crust, its oceans, or its atmosphere. Define each term (geology, oceanography, and meteorology) for students.
9. Have the students tape their brainstorming slips on the appropriate sheet of chart paper. (Students will probably have to break up the groupings they made earlier.) Then have them record the chart groupings on their record sheets.
10. Discuss the results. Lead students to recognize how Earth's components are grouped into major categories.

Name ______________________________

Earth's Components

Procedure and Observations

1. Imagine you are someplace on Earth. What natural things do you see, hear, smell, feel, and taste? Jot down some ideas.

____

____

____

2. Sort your idea slips into three groups using the chart below. Then come up with a title for each group.

Group 1	Group 2	Group 3
Items:	**Items:**	**Items:**
Title:	**Title:**	**Title:**

3. Now sort the same slips into the following three groups.

Geology	Oceanography	Meteorology
Items:	**Items:**	**Items:**

Investigation 2

Looking at Rocks

Materials

See advance preparation on page 2.

- student record sheet on page 9, reproduced for each student
- overhead transparency of *Rock Types* on page 3
- magnifiers
- rocks, assorted and numbered

Steps to Follow

1. Divide the class into small groups. Have each group play the game "I Spy with My Little Eye" to practice describing the properties of an object. (Students mentally identify an object in the classroom and then describe its properties to the rest of their group. The group uses the clues to identify the object.)
2. Discuss with students what kinds of clues worked best. They should recognize that very few objects can be identified with just one or two clues.
3. Distribute a collection of rocks and some magnifiers to each group. Encourage students to examine the rocks and record their observations of the rocks' properties on their record sheets.
4. Once students have finished making their observations, have them share their results in small groups or as a class. Challenge the other students to identify the rock being described by looking at their own notes.
5. Using the *Rock Types* transparency, discuss how the three different types of rock (igneous, sedimentary, and metamorphic) are formed.
6. Challenge students to guess which of their rocks are igneous, which are sedimentary, and which are metamorphic, based on their properties. Have students record their guesses in the last column of their observation charts.

Follow-Up

Introduce the concept of the rock cycle. Using the *Rock Types* transparency, add arrows and labels that show how each type of rock can change to another type.

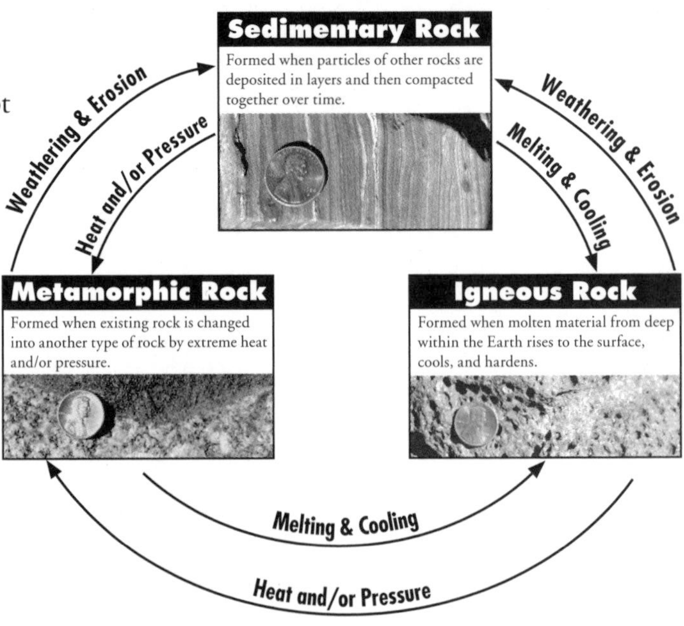

Name ____________________________________

Looking at Rocks

Procedure and Observations

1. Use the magnifiers to observe the rock samples. Record the properties of each specimen on the chart below.

Sample #	Color	Density	Texture	Other	Origin
1					
2					
3					
4					
5					
6					
7					
8					
9					
10					

2. Which rocks do you think are igneous? Sedimentary? Metamorphic? Use the properties you described above to guess the origin of each rock sample. Record your guesses in the last column of the chart.

Conclusion

3. In what ways were the rocks alike? In what ways were they different?

____

____

____

Investigation 3

Looking at Minerals

Materials

See advance preparation on page 2.

- student record sheet on page 11, reproduced for each student
- overhead transparency of *Mineral Properties Chart* on page 4
- glass plates (for scratch tests)
- magnifiers
- minerals, assorted and numbered
- pennies (for scratch tests)
- steel nails (for scratch tests)

Steps to Follow

1. Distribute a collection of minerals and some magnifiers to each group. Encourage students to examine the minerals and record their observations of the minerals' properties on their record sheets.
2. Once students have finished making their observations, have them share their results in small groups or as a class. Challenge the other students to identify the mineral being described by looking at their own observation notes.
3. Using the *Mineral Properties Chart* transparency, explain that Mohs' Scale of Hardness is a scale that tells how hard each mineral is. Copy the following chart on the board.

Mohs' Scale Number	Common Tests
1	scratched by fingernail
2	
3	scratched by copper penny
4	scratched by steel nail
5	
6	scratches glass
7	
8	
9	
10	scratches all common materials

4. Have students use the information provided in the transparency to try to identify the minerals they observed. Provide guidance as needed.

Follow-Up

Have students research how people and industry use each of the minerals they observed. For example, gypsum is used to make plaster of Paris, and galena is used in batteries and paints. They may also want to research where each mineral is found.

Name ______________________________

Looking at Minerals

Procedure and Observations

1. Use the magnifiers to observe the mineral samples. Record the properties of each specimen on the chart below.

2. Use the information on the transparency to try to identify each of the mineral specimens. Record each mineral's name in the last column of the chart.

Sample #	Color	Density	Luster	Hardness	Name
1					
2					
3					
4					
5					
6					
7					
8					
9					
10					

Conclusion

3. How are we able to tell minerals apart?

____

____

____

Investigation 4

Ocean Water v. Fresh Water

Materials

See advance preparation on page 2.

- student record sheet on page 13, reproduced for each student
- balances or scales
- fresh water (collected or spring water)
- graduated cylinders or beakers
- ocean (salt) water (collected or prepared)
- clear plastic cups
- penlights
- petri dishes
- white paper

Steps to Follow

1. Divide students into small groups. Distribute a container of ocean water and a container of fresh water to each group.
2. Have the students compare and contrast the two samples of water. Some comparison ideas include the following:

   Density – Measure the mass of equal amounts of each sample. Use a graduated cylinder or beaker to measure equal amounts. Use a balance or scale to find the mass of each sample.

   Clarity – Compare the amount of light that passes through each sample. Use a penlight to shine a beam of light through each sample onto a white piece of paper. Compare the brightness of the light that passes through each sample.

   Dissolved Material – Place equal amounts of ocean water and fresh water into petri dishes and allow the water to evaporate overnight (or over several days). Examine what is left behind in each dish.

3. Encourage students to enter all experimental data on their record sheets.

Follow-Up

Have the students research various animals that live in saltwater and freshwater environments. How do saltwater animals deal with the excess salt in the water?

Name ____________________________________

Ocean Water v. Fresh Water

Investigation 4

Prediction

1. What differences do you think you might find between ocean water and fresh water?

____

Procedure and Observations

2. Perform different tests on the samples of ocean water and fresh water.

Density: Measure out equal amounts (in cm^3) of each type of water. Place each sample on a balance. Measure the mass of each sample in grams (g). Record the data on the chart below. Then calculate the density of each by dividing mass by volume.

	Volume (cm^3)	Mass (g)	Density (g/cm^3)
Ocean Water			
Fresh Water			

Clarity: Fill a glass or clear plastic cup with ocean water. Fill a second glass or cup with fresh water. Holding a penlight on one side of the glass, shine the light through the water so that it strikes a piece of white paper held up against the other side of the glass. Compare the brightness of the two spots of light. What did you observe?

____

Dissolved Materials: Pour a small amount of ocean water into one half of a petri dish, or other shallow dish. Pour an equal amount of fresh water into a second dish. Leave them overnight and observe the dishes the next day. Do you see anything in the dishes?

____

3. Perform any other tests you can think of to compare the two types of water. Record your observations on the back of this sheet.

Conclusion

4. What differences did you find between ocean water and fresh water?

____

____

Investigation 5

Layers of the Atmosphere

Materials

- student record sheets on pages 15–17, reproduced for each student
- overhead transparency of *Layers of the Atmosphere* on page 5
- assorted materials for atmosphere models (shoeboxes, string, empty soda bottles, colored sand, construction paper, etc.)

Steps to Follow

1. Show students the *Layers of the Atmosphere* transparency. Remind students that the atmosphere is the layer of gases and particles that surrounds Earth. Among other things, it contains oxygen, ozone, and water vapor (water in its gaseous state). Explain that scientists divide up the atmosphere into layers based mainly on which gases are found where.
2. Have students use the information presented in the transparency to create a diagram of the atmospheric layers on their record sheets.
3. Now have students use the temperature data on their record sheets to graph the temperature of the atmosphere as it changes with altitude. Help them as needed with the graphing.

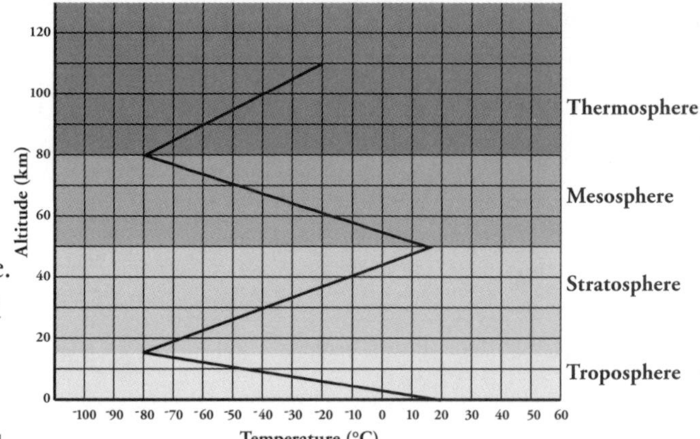

4. Once students have completed their graphs, discuss their results. (Students will find that the temperature changes between cold and warm as they move through different layers of the atmosphere. This is because some layers contain gases that absorb solar energy, and others do not.)
5. Divide students into small groups. Challenge the groups to come up with physical models for the layers of the atmosphere. They may choose to use a shoebox with string pulled across it to represent different altitudes. Or they may choose to fill an empty soda bottle with layers of colored sand. Encourage both creativity and accuracy. Tell students to plan their designs before constructing them. Have students record their ideas and draw their completed models on their record sheets.

Follow-Up

Have students research air pressure and how it changes with altitude. How does the pressure data graph compare to the temperature data graph within the troposphere? (Both pressure and temperature decrease with altitude.)

Name ________________________________

Investigation 5

Layers of the Atmosphere

Procedure and Observations

1. Use the information presented in the transparency to make a diagram of the layers of the atmosphere. Be sure to include altitude data in the diagram.

Altitude (km)

Temperature (°C)

2. The data table below shows the average temperature at different levels in Earth's atmosphere. Use the data to create a graph that shows what happens to temperature as we pass from one layer to another.

Altitude Above Earth's Surface (km)	Temperature (°C)
0	20
16	-80
50	15
80	-80
110	-20

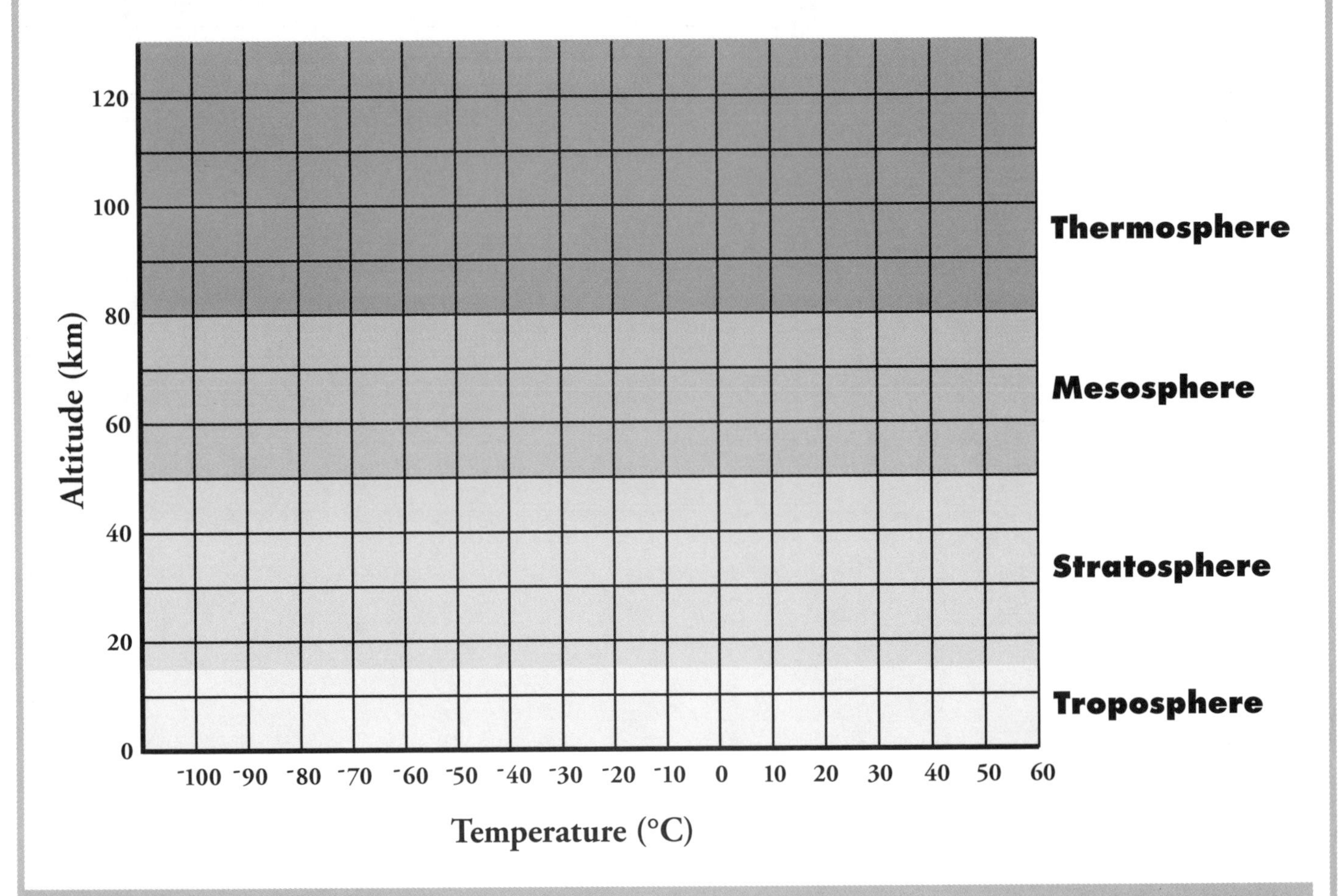

Conclusion

3. Based on your graph, what can you conclude about how temperature changes as you move farther up in Earth's atmosphere?

____

Procedure and Observations

4. Work with your teammates to design a model you will build to represent the layers of the atmosphere. Think about what materials you will use and what information your model will show. Describe your model in the space below.

5. Gather your materials and construct your model. Draw a picture of what your model looks like in the space below.

2

The core, mantle, and crust make up the layers of the Earth.

Prepare in Advance

Investigation 1: Prepare one hard-boiled egg for each pair of students. Boil the eggs for 15 to 20 minutes. Do not peel.

Teacher Information

At this grade level, we tend to describe Earth as being made up of three distinct layers: the **crust,** the **mantle,** and the **core.** Although nobody has actually seen any of the inner layers, evidence indicates that below the very thin crust lies a molten mantle very similar in composition to the crust, but in a molten state. Below the mantle and at the very center of the Earth lies the core. Composed mainly of crystalline nickel, the core is in a solid state. Although it is even hotter than the mantle, extreme pressure causes the core to be a solid.

When Earth was first forming, it was molten throughout. Gravity caused heavier materials, such as iron and nickel, to sink to the center of the Earth. Lighter materials floated to the surface. As Earth began to cool, these lighter materials formed a hard crust across the surface.

Although it seems solid, Earth's crust is actually broken into several large pieces, called **tectonic plates** (or **crustal plates**). Evidence suggests that **convection currents** within the mantle move these plates in different directions. These currents are the result of hot mantle material rising toward the crust, cooling, becoming more dense, and sinking once again toward the center of the Earth. As the molten material moves, it drags the plates sideways along Earth's surface.

In general, three different situations can be observed at the junction of two plates:

- The plates can crash together, causing mountains to form. (For example, the Himalaya Mountains formed as the Indian plate crashed into the Asian plate.) Or, one plate can submerge under another plate. This action often results in volcanic activity.
- The plates can separate from each other, allowing magma to well up from below and form new land between the plates. For example, the mid-Atlantic ridge formed this way.
- The plates can move sideways against each other, creating major earthquakes. The San Andreas earthquake zone in California is the location of such a junction.

The **Ring of Fire** describes a series of active volcanoes (and seismic activity) that line the Pacific Ocean. When the locations of these volcanoes are overlaid on a map of the plate boundaries, one can see that the two coincide. Where the plates are separating, magma is able to travel up through the crust, causing volcanic action. Where the plates are scraping against one another, there is seismic activity.

Layers of the Earth

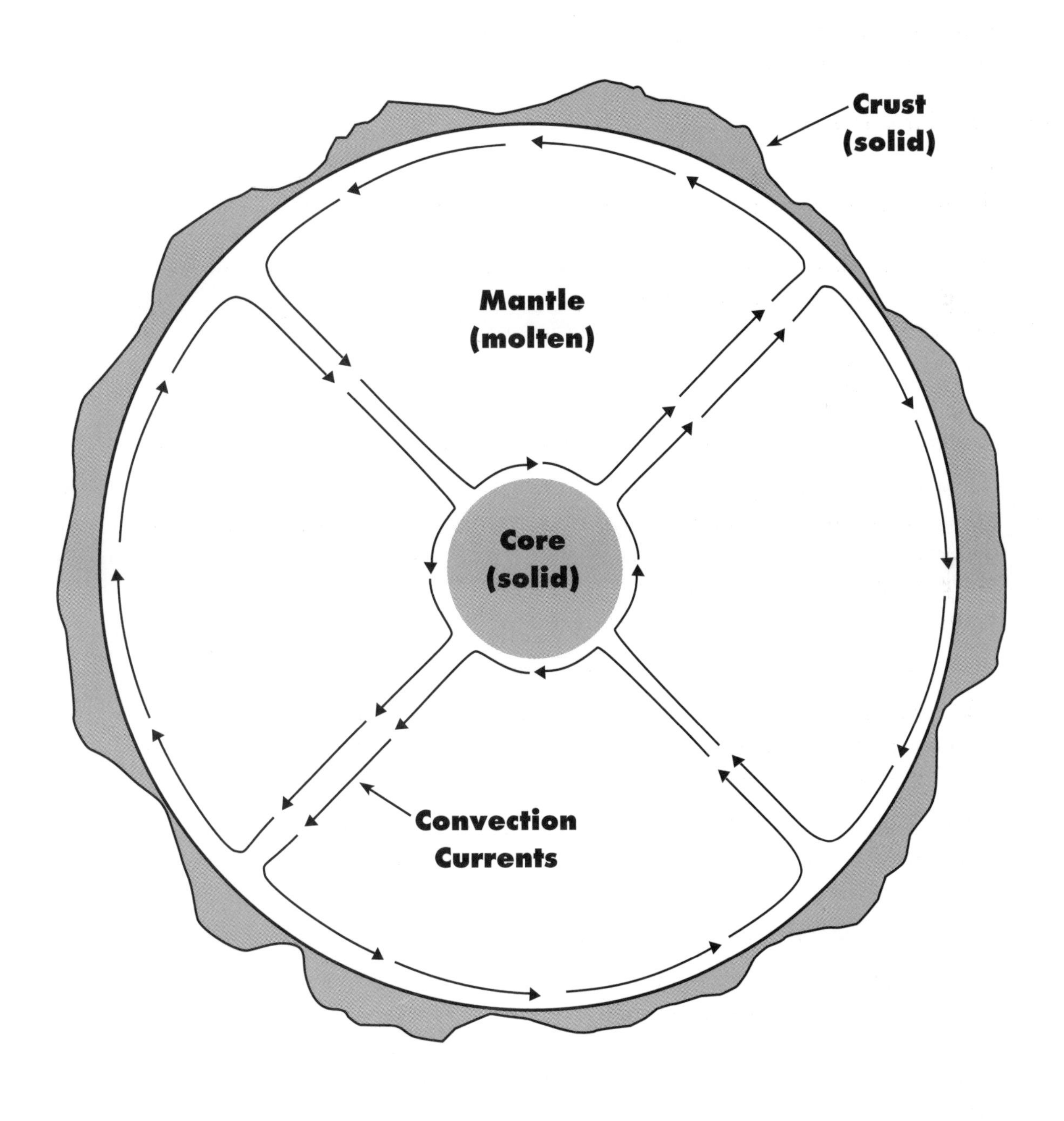

Concept
2
Planet Earth

World Map

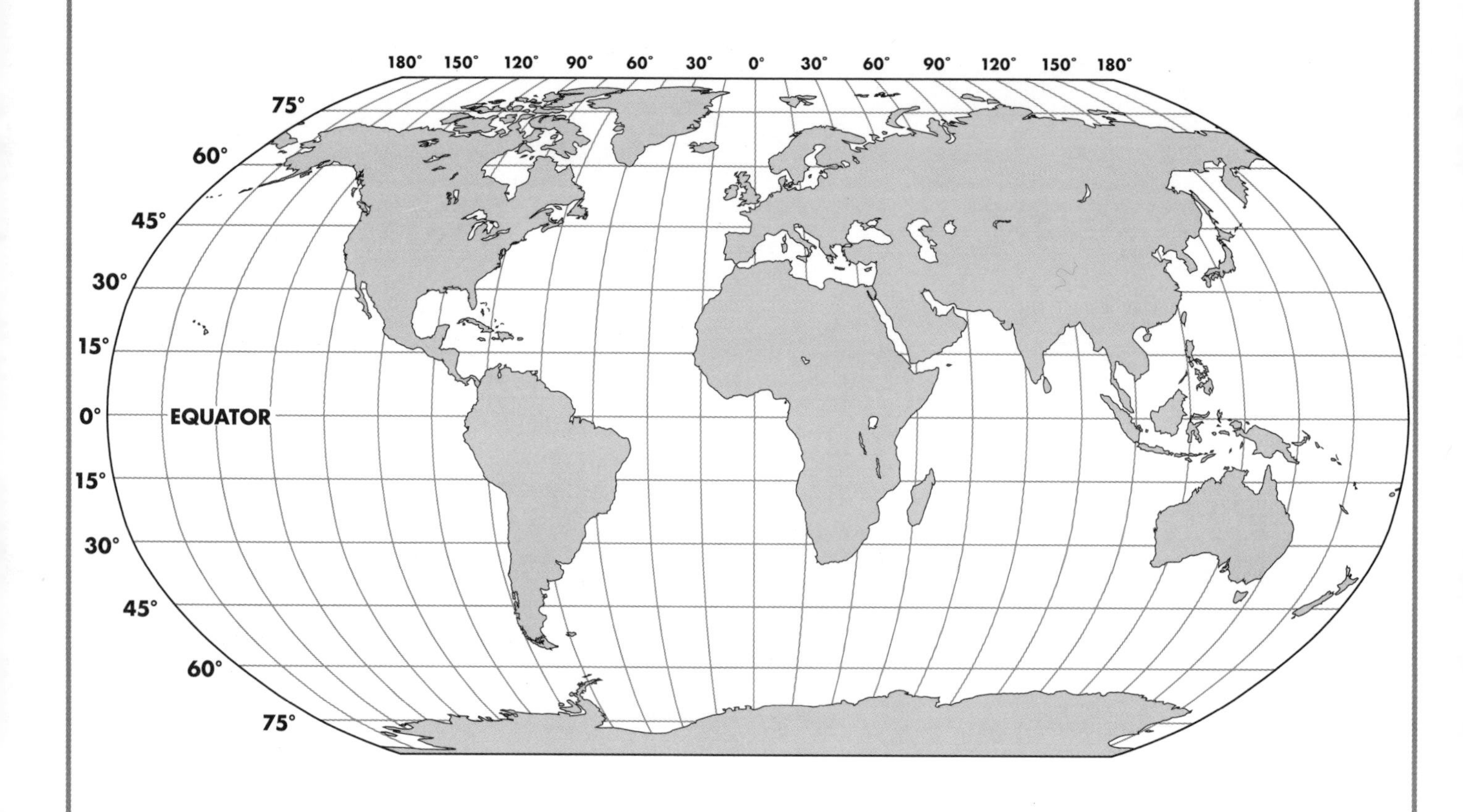
180° 150° 120° 90° 60° 30° 0° 30° 60° 90° 120° 150° 180°
75°
60°
45°
30°
15°
0°
EQUATOR
15°
30°
45°
60°
75°

Crustal Plate Boundaries

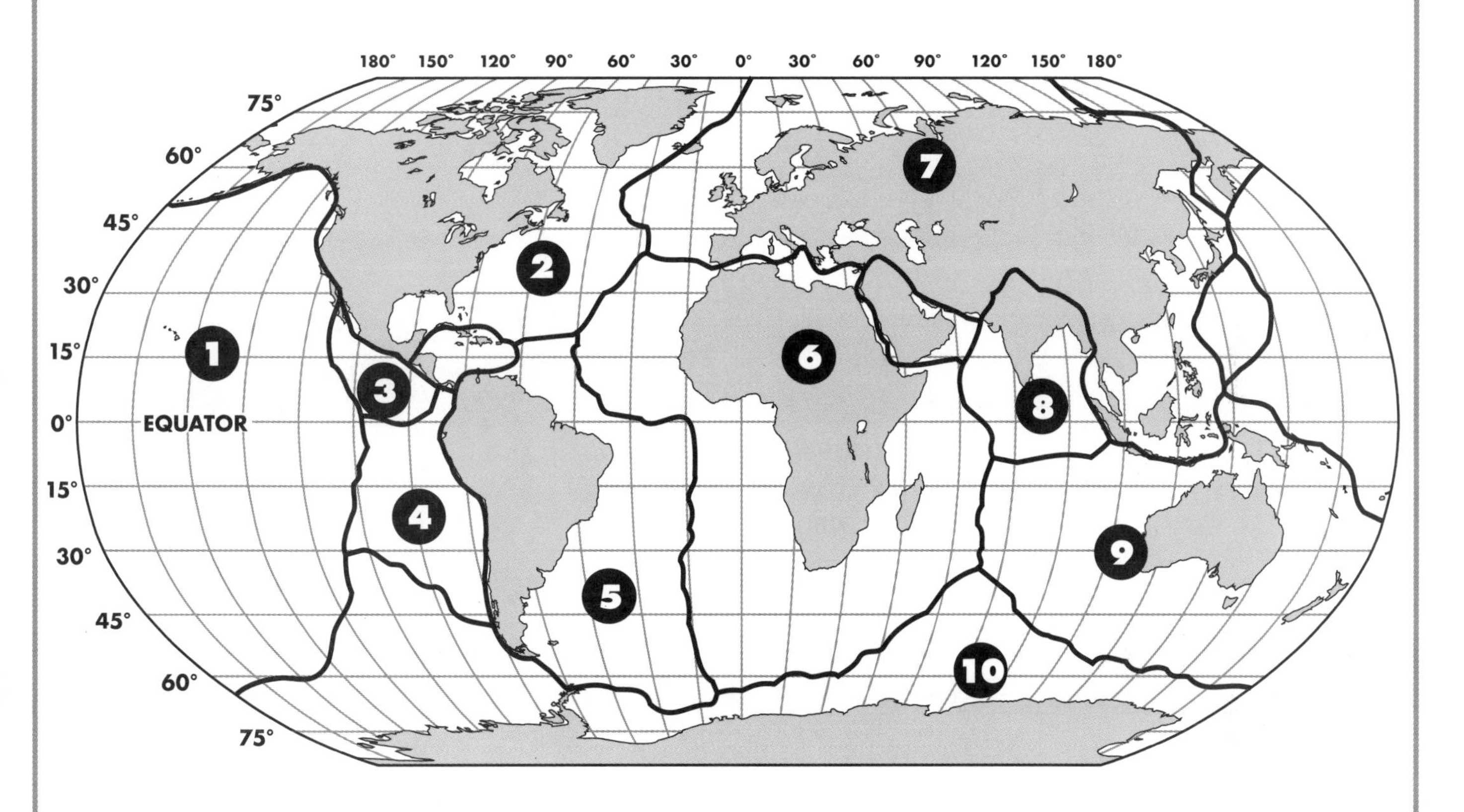

1. **Pacific plate**
2. **North American plate**
3. **Cocos plate**
4. **Nazca plate**
5. **South American plate**
6. **African plate**
7. **Eurasian plate**
8. **Indian plate**
9. **Australian plate**
10. **Antarctic plate**

Investigation 1

Layers of the Earth

Materials

See advance preparation on page 18.

- student record sheet on page 23, reproduced for each student
- overhead transparency of *Layers of the Earth* on page 19
- hard-boiled egg for each pair of students
- dull knives

Steps to Follow

1. Pass out an egg to each pair of students. ***Caution: Explain to the students that the egg is not for eating. Warn students never to eat in science class unless told to do so by their teacher.***
2. Have students cut the egg in half through the shell. (You may choose to do this step yourself and distribute the cut eggs.)
3. Have the students describe the makeup of the egg, paying particular attention to its multiple layers. Have them record their observations on their record sheets.

4. Show students the *Layers of the Earth* transparency. Go over each of the terms.
5. On their record sheets, have students describe how their model is similar to the composition of Earth.
6. Discuss students' observations with them. Point out that, like their egg, the Earth is made up of three main layers. Both the Earth and the egg have a hard, thin outer shell. This layer of the Earth is called the **crust**. Earth's crust is made up of rock. The white part of the egg roughly corresponds to Earth's **mantle**, except that Earth's mantle is molten, not solid. Like the yolk, Earth's **core** is solid. Intense heat in Earth's core warms the rock of the mantle, making it molten, or liquid.
7. Once students have completed their observations, have them throw away their egg pieces and wash their hands thoroughly with soap and water.

Follow-Up

Challenge students to do some research on the different layers of the Earth and come up with a more accurate model.

Name ____________________

Concept 2 Planet Earth

Investigation 1

Layers of the Earth

Procedure and Observations

1. Draw and describe what a cross section of your egg looks like.

2. Draw and describe what a cross section of the Earth looks like.

Conclusion

3. How is your egg cross section similar to the cross section of Earth? In what ways is it different?

Investigation 2

Crustal Plates

Materials

- student record sheet on page 25, reproduced for each student
- *World Map* on page 20, reproduced for each student
- overhead transparency of *Crustal Plate Boundaries* on page 21
- pencils and pens

Steps to Follow

1. Remind students that the Earth is enclosed by a thin, hard crust. Explain that Earth's crust is actually broken into pieces called **crustal plates,** or tectonic plates. These plates are constantly moving around, bumping into one another in some places and separating in others.
2. Show students the *Crustal Plate Boundaries* transparency. Have them note that the continents are part of the plates, but are not themselves the actual plates. Point out that most plates contain both continents and oceans.
3. Have students record the names of the major crustal plates on their record sheets.
4. Ask students what can be found at the junction of the North American plate and the Pacific plate (the Sierra Nevada). Ask students to speculate as to the movement of the two plates at that point. (They are moving toward each other, bumping into one another and forming a range of mountains.)
5. Distribute a copy of the *World Map* to each student. Students will also need pencils and pens. Have students draw in the crustal plate boundaries on their World Maps.
6. Students will use their completed maps again in Investigation 4. You may choose to collect all the maps and distribute them again later.

Follow-Up

Make photocopies of the *Crustal Plate Boundaries* sheet. Give each student a copy and have them cut out the pieces. Using the *Crustal Plate Boundaries* transparency as a guide, challenge students to put together the pieces like a jigsaw puzzle.

Name ____________________

Crustal Plates

Procedure and Observations

1. In the space below, record the names of the major crustal plates.

2. Using the overhead transparency as a guide, draw the crustal plate boundaries on the World Map your teacher gives you. Use a pencil at first. Once you've finished drawing the boundaries, go back over the lines with dark ink.

Conclusion

3. Based on what you know about the crustal plates, what can you infer about the Andes Mountains along the western coast of South America?

Investigation 3

Convection Currents

Materials

- student record sheet on page 27, reproduced for each student
- overhead transparency of *Layers of the Earth* on page 19
- 1-liter glass beaker (Pyrex)
- glitter (a dark color)
- hot plate

Steps to Follow

1. Perform the following demonstration for the class:

   Preheat the hot plate to a medium-high setting. Make sure the glass beaker can be heated on a hot plate. Fill the beaker ¾ full of water. Sprinkle some glitter on top of the water. Have students record their observations on their record sheets.

   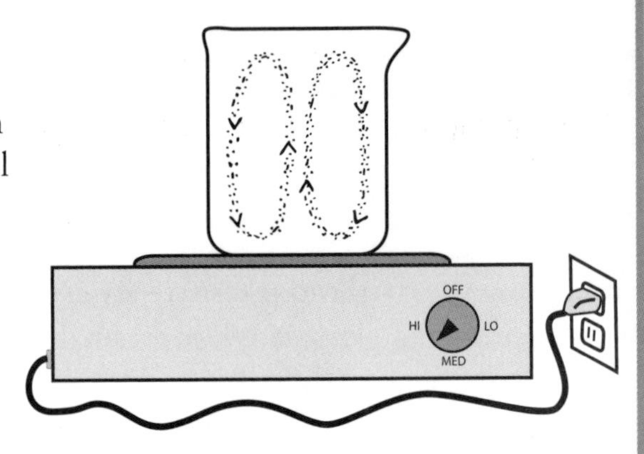

   Place the beaker onto the hot plate. Once the water begins to heat up, convection currents within the water will pull the glitter in a circular path from the top of the beaker to the bottom. The glitter allows students to see how the water moves in circular currents throughout the beaker. Have students record their observations on their record sheets.

2. Once students have completed their observations, hold a class discussion. Begin by reminding students that Earth's crust is broken up into crustal plates, and that these plates are constantly moving around.

3. Show students the *Layers of the Earth* transparency. Ask students to describe how the model they created today is related to the "convection currents" labeled on the transparency. (Like the water in the beaker, molten rock moves in circular paths through the mantle.)

4. Have students describe what they think makes crustal plates move around. (Earth's hot core warms the liquid mantle, much as the hot plate warms the water in the beaker. The crustal plates are pulled sideways along the liquid's surface as the liquid moves in a circular path from the top of the mantle, to the core, to the top of the mantle again.)

Follow-Up

Have students research the makeup of the mantle. What kinds of rock material are found there? What is the average temperature?

Name ____________________

Convection Currents

Concept 2 Planet Earth

Investigation 3

Procedure and Observations

1. Observe the beaker of water before and after it is heated on the hot plate.
2. What happened when your teacher first added the glitter?

____

____

3. What happened as the water heated up? Draw and describe what you saw.

____

____

____

Conclusions

4. How does this model demonstrate what actually goes on inside Earth's mantle?

____

____

____

5. What moves Earth's crustal plates?

____

____

____

Investigation 4

Ring of Fire

Materials

- student record sheet on page 29, reproduced for each student
- overhead transparency of *Crustal Plate Boundaries* on page 21
- colored grease pencils or erasable markers
- colored pencils (red and blue)
- completed maps from Investigation 2

Steps to Follow

1. Talk to the students about maps and latitude and longitude. Review unfamiliar terms and concepts as needed.
2. Distribute students' maps from Investigation 2. Each map should show the continents as well as the locations of the plate boundaries.
3. Photocopy the data below and distribute a copy to each student.
4. Have students plot each of the earthquakes on their map using a blue dot. Then have students plot each of the volcanoes on their map using a red dot.
5. With the projector turned off, plot the locations of the earthquakes and volcanoes on the *Crustal Plate Boundaries* transparency using colored grease pencils or erasable markers.
6. When students have plotted all of their dots, have them look for any patterns in the maps. (Much of the volcanic and seismic activity is centered along the edge of the Pacific plate, hence the name "Ring of Fire." The Hawaiian Islands are one notable exception.) Have the students discuss any patterns that they see.
7. Display the *Crustal Plate Boundaries* transparency with colored dots. Lead students to conclude that where plates are moving against one another or away from one another, lots of activities (like earthquakes and volcanic eruptions) are taking place.

Earthquakes

	Latitude	Longitude
1	30° N	118° W
2	65° N	15° W
3	33° N	30° W
4	5° S	152° E
5	27° N	105° W
6	60° N	145° W
7	30° N	75° E
8	35° N	135° E
9	30° S	100° W
10	30° S	10° W
11	32° N	30° E
12	58° S	60° W

Volcanoes

	Latitude	Longitude
1	35° N	135° E
2	40° N	125° W
3	42° N	110° E
4	5° S	130° E
5	60° N	140° W
6	40° N	30° E

Name ________________________________

Concept 2 Planet Earth

Investigation 4

Ring of Fire

Procedure and Observations

1. Use the latitude and longitude data your teacher gives you to plot the locations of earthquakes and volcanoes on your map.

2. What patterns do you notice when you look at the location of the dots on your map?

____

____

____

____

Conclusions

3. What conclusions can you draw about crustal plates and major Earth activity based on what you see in your map?

____

____

____

____

____

____

4. Where do you suppose the name "Ring of Fire" came from?

____

____

____

____

3

Earth's crust contains continents and oceans.

Teacher Information

The surface of the Earth is divided between land and water—continents and oceans. Oceans cover almost three-quarters of Earth's surface. Continents make up the rest. Earth's surface as it appears today is not how it appeared in the past. As crustal plates are pushed and pulled, continents are moved around the globe. Millions of years ago, all the continents we see today were grouped together in a giant mass known as **Pangaea.** Looking at the continents today, we can see how they might have fit together at one time. (See art on page 34.)

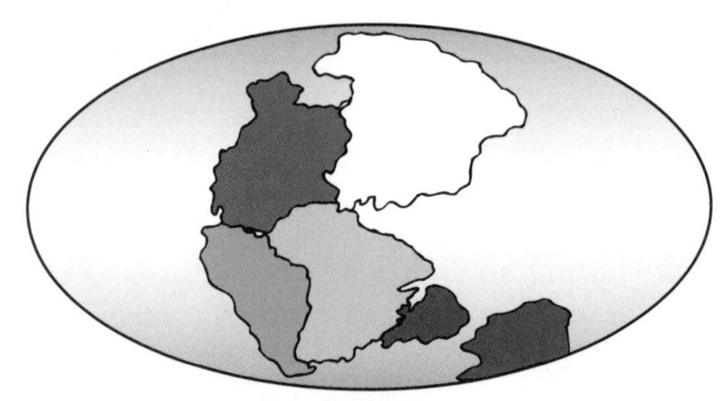

About 245 million years ago

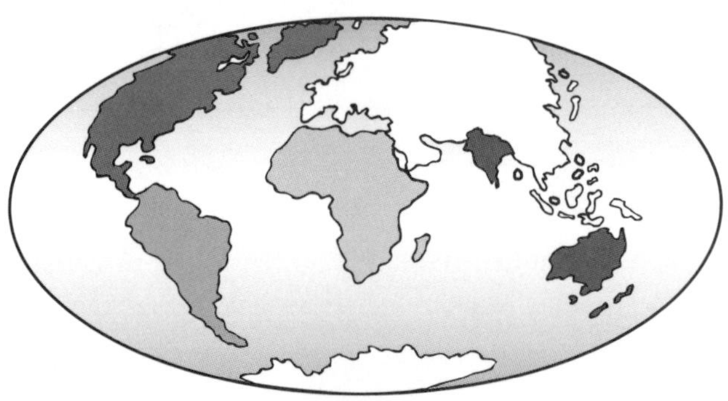

Today

Just as the continents show great variety in landforms, so does the ocean floor. Mountain ranges, plains, and deep trenches make up an underwater landscape that is every bit as diverse as that seen on land. There are two major regions of ocean floor—the region associated with the continental shelf (and therefore shallower) and the region associated with the oceanic crust (and therefore deeper). Most of the major landforms, including plains, ridges, trenches, valleys, and seamounts, are found on the oceanic crust.

Creating an accurate map of Earth's surface is not as easy as it seems. Transferring a curved surface to a flat map is an imperfect process. Some distortion exists with any map projection. One of the most familiar maps we use is the **Mercator projection.** The Mercator projection map has straight lines of longitude and latitude that intersect at right angles. The projection shows zero distortion within 20 degrees of the equator, but shows greater and greater area distortions as you move toward the poles. On a Mercator projection, Greenland looks as large as Africa, when in reality Africa is 15 times larger. The Mercator projection is often used for marine navigation because shorelines are represented more accurately than they appear on other maps. Also, sailors can navigate simply by drawing a straight line between two points and then following the compass direction on the map. Other maps carry their own distortions.

Landmass Shapes

Concept
3
Planet Earth

Sea Floor

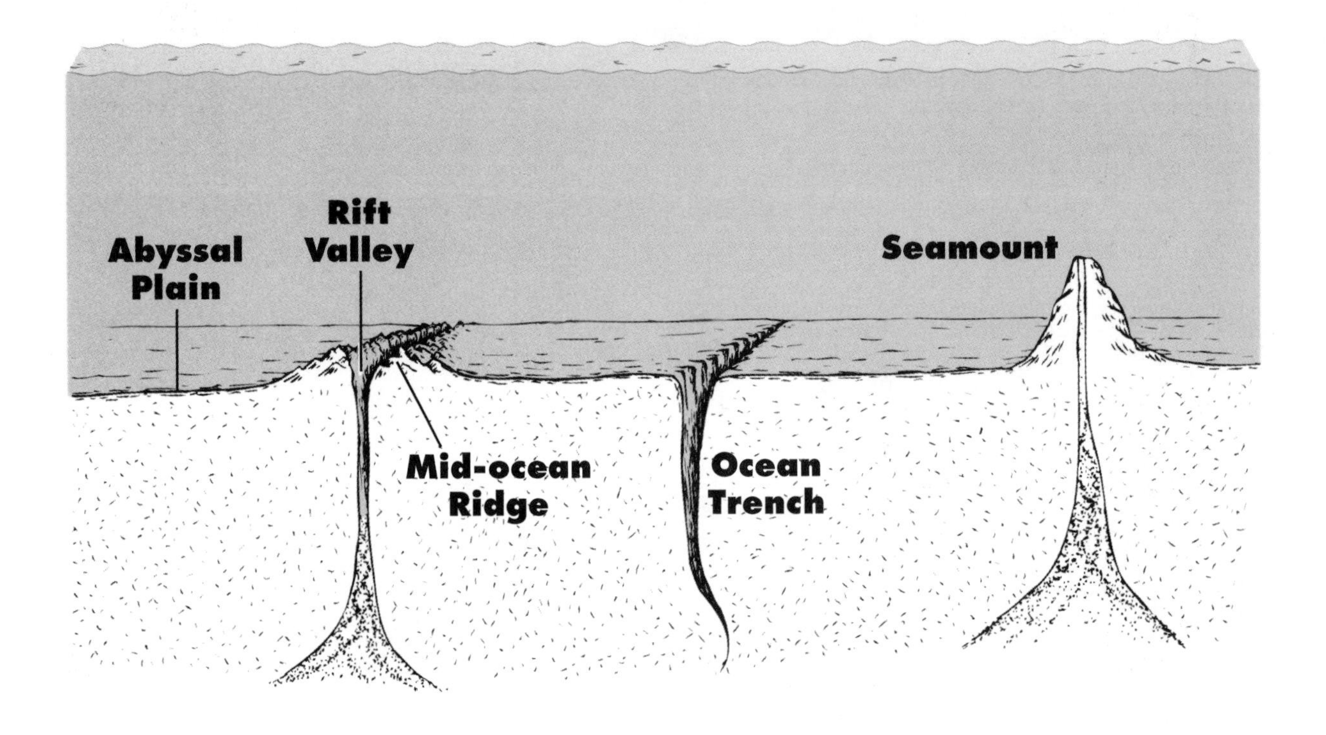
Rift Valley
Abyssal Plain
Seamount
Mid-ocean Ridge
Ocean Trench

Mercator Projection

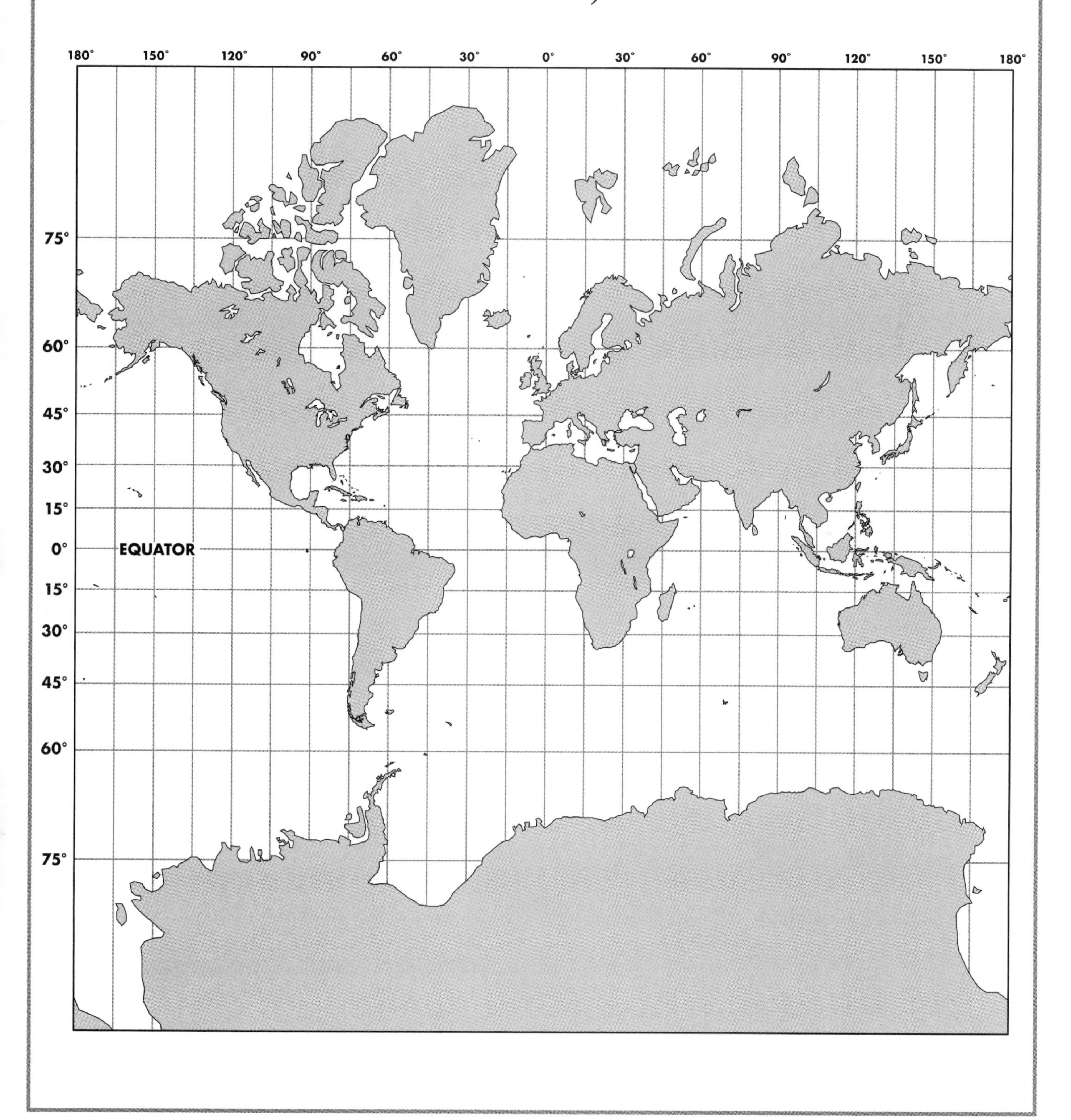

Investigation 1

Pieces of a Puzzle

Materials

- student record sheet on page 35, reproduced for each student
- *Landmass Shapes* on page 31, reproduced for each student
- scissors
- tape

Steps to Follow

1. Distribute a copy of the *Landmass Shapes* to each student. Remind students that these shapes show six of the Earth's seven major continents and India; Antarctica has been omitted.
2. Have students cut out each continent along the coastline. Have them place the pieces on their desk and move them around a bit. Ask them if they notice anything interesting.
3. Borrowing one student's pieces, show how Africa and South America seem to fit together.
4. Have students try to fit together other shapes.
5. Introduce the term **Pangaea** to students. Explain that scientists use this term to describe the single continent that they believe existed millions of years ago on Earth's surface. Scientists believe that, over time, Pangaea broke apart and each of the pieces drifted into its present place. According to this theory, the continents are still drifting. (You may want to show students the pictures on page 30 that show how Pangaea broke apart.)

Follow-Up

As a review for Concept 2, challenge students to explain how these continents could have drifted (through the action of convection currents in the mantle).

Name ______________________________

Pieces of a Puzzle

Procedure and Observations

1. Cut out the continents. Move the pieces around on your desk.

2. Do you notice anything unusual about how the shapes fit together?

____

____

____

3. Once you have fit your pieces together, draw a picture of them in the space below.

Conclusion

4. Why do many of the continents appear to fit together?

____

____

____

____

Investigation 2

Sea-floor Model

Materials

- student record sheet on page 37, reproduced for each student
- overhead transparency of *Sea Floor* on page 32
- clay (various colors)
- shoeboxes

Steps to Follow

1. Ask students what they think the bottom of the sea looks like. Have students record their ideas on their record sheets. Ask volunteers to draw some of their ideas on the board. (Most students do not realize that the ocean floor is as varied as the land they see above the water.)
2. Display the *Sea Floor* transparency. Have students identify the various landforms shown (plains, valleys, trenches, ridges, and seamounts). Review how each landform is created.
   - **Mid-ocean ridges** form where crustal plates move away from one another, creating a gap through which magma rises up. The hot magma causes the crust on either side of the rift to swell, forming mountains.
   - A **rift valley** forms between the mountains on either side of the rift, or gap, in the sea floor.
   - An **ocean trench** forms where one crustal plate is pushed under another crustal plate.
   - **Seamounts** form where magma pushes its way up to create a bulge. If the seamount rises above sea level, it becomes a volcanic island.
   - The flat portion of the ocean basin is called the **abyssal plain**. Here the remains of dead organisms pile up along with mud and other debris.
3. Lead students to understand that the same forces that created landforms on the continents (movement of crustal plates, volcanic activity, and earthquakes) created landforms on the sea floor.
4. Divide the class into small groups. Distribute boxes and clay to each group. Challenge students to first design, and then build, a sea-floor model based on what they learned from the transparency.

Follow-Up

If you have a ceramics lab in your school, you may want to have students glaze and fire their model landforms before adding them to the bottom of an actual aquarium.

Name ______________________________

Concept 3 Planet Earth

Investigation 2

Sea-floor Model

Procedure and Observations

1. What do you think the sea floor looks like? Draw your ideas below.

2. Look at the *Sea Floor* transparency. What features does it contain that you did not imagine were there?

____

____

____

3. Use the clay your teacher gives you to create a model of the sea floor with the other members of your group. First create a plan and agree on it as a group. Then begin building your model, giving each group member the chance to create one landform. Draw your completed model on the back of this sheet.

Investigation 3

Map Projections

Materials

- student record sheet on page 39, reproduced for each student
- overhead transparency of *Mercator Projection* on page 33
- classroom globe

Steps to Follow

1. Place the globe in front of the class. Review with students the major continents, and how to read lines of latitude and longitude.
2. Ask students how they would make a flat map of the globe. Have students offer ideas while you point out the disadvantages of each arrangement as far as distortion of size, direction, and distance. Lead students to conclude that there is no perfect flat map of planet Earth. Certain maps can show some parts accurately, but they will always contain some distortions because the Earth is a sphere, not flat.
3. Using the *Mercator Projection* transparency, explain that the Mercator projection was developed by the Flemish (what is now Belgium) geographer Gerardus Mercator in 1569. Explain that the Mercator projection is probably the most widely recognized map in the world.
4. Go over the major features of the map with students. (The lines of longitude are perpendicular to the lines of latitude. The lines of longitude are evenly spaced, but the lines of latitude are not.)
5. Ask students to compare the map to the globe and decide which elements are accurately represented and which are not. (The map shows the equatorial regions accurately, but distorts size near the poles. The wider gaps between the latitude lines near the poles are meant to address this problem of size exaggeration.) Have students record their observations on their record sheets.
6. Explain that although the Mercator projection shows size distortions near the poles, it still is considered one of the most useful projections because sailors can use it to navigate the ocean easily. A sailor simply draws a line between two points (where the ship is and the desired destination) and follows the compass direction indicated by the map.

Follow-Up

Have students research other map projections (conic projections, azimuthal projections) and identify their advantages and disadvantages. Discuss the fact that some projections work well for one task but not for another. While no one map may be perfect, each map serves a particular function.

Name ______________________________

Map Projections

Procedure and Observations

1. Look at the *Mercator Projection* transparency.

2. In what ways is the Mercator projection an accurate map? In what ways is it not?

Ways the Mercator Projection Is Accurate	Ways the Mercator Projection Is Not Accurate

Conclusion

3. Why can't we make a flat map of Earth that is totally accurate?

____

____

____

____

____

____

4

Constructive forces affect landforms.

Prepare in Advance

Investigation 3: Mix powdered clay in enough water to make a cloudy mixture.

Teacher Information

Earth's crust is forever changing. New material is added while other material is moved around or engulfed by the mantle. Some constructive forces acting on the Earth include volcanic eruptions, mountain building, and **deposition.** Each of these actions adds new material to Earth's crust.

Volcanoes are the link between the Earth's mantle and its crust. Volcanoes form where magma flows upward from the mantle and pours out onto the crust. Once on the surface of the crust, the hot magma cools to form **lava.** In this way, new rock material is added to the surface of the Earth.

Extrusive igneous rocks are rocks that have solidified from molten magma at or near the Earth's surface. The Hawaiian Islands, an excellent example of volcanic islands, are made up entirely of extrusive igneous rocks. **Intrusive** igneous rocks, by comparison, form from molten magma that cools slowly somewhere beneath Earth's surface. Intrusive crystals are generally larger than those of extrusive igneous rock.

Recall that Earth's crust is made up of a series of pieces called **tectonic plates** (or **crustal plates**). Convection currents in the mantle move portions of the crust in different directions. Where two plates collide, the plates can buckle and mountains can form. For example, the Himalayas formed as the Indian plate crashed into the Asian plate.

Erosion is the process whereby small pieces of soil, rock, and other material are carried by wind or water from one place to another. Rivers and streams often carry large amounts of silt and sand to larger bodies of water (lakes and oceans) where the debris is deposited. As particles settle out, they form new land. In the case of a river delta, deposition creates new land along a coastline.

Three Types of Volcanoes

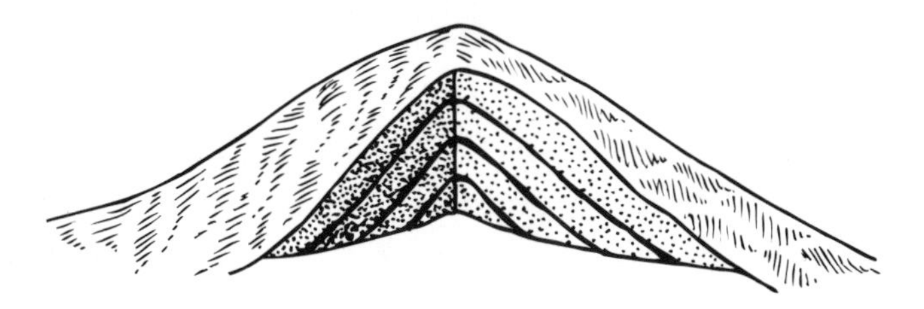

Shield Volcano

Built from layers of lava that flowed slowly out of the Earth, the shield volcano's sides are not very steep.

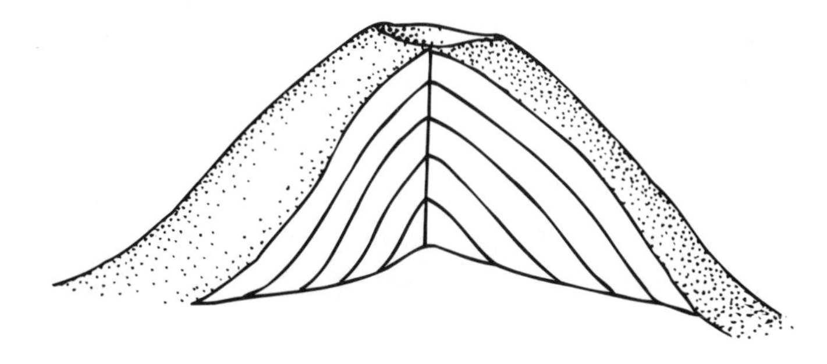

Cinder Cone Volcano

Explosive eruptions shoot magma out of the cinder cone volcano where it falls as layers of stone and ash.

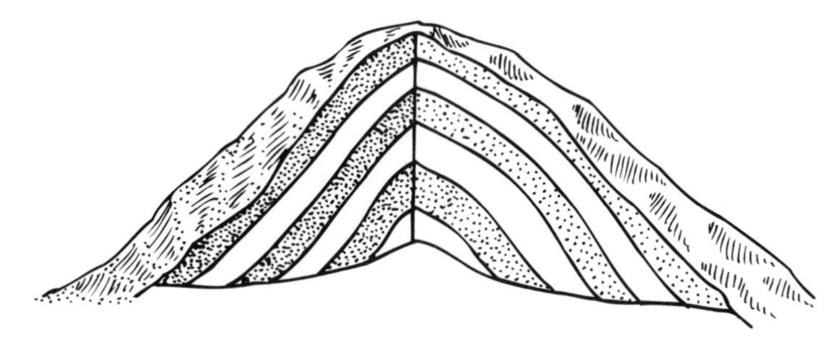

Composite Volcano

Powerful eruptions followed by gentle oozing of lava create alternating layers of lava and stone/ash in the composite volcano.

Investigation 1

Build a Volcano

Materials

- student record sheet on page 43, reproduced for each student
- overhead transparency of *Three Types of Volcanoes* on page 41
- 18-inch (45.5 cm) cardboard square
- plastic soda bottle
- hot glue
- newspaper
- heavy-duty aluminum foil
- flat black spray paint
- spray adhesive
- dirt
- teaspoon
- tablespoon
- red food coloring
- dishwashing liquid
- baking soda
- water
- vinegar

Steps to Follow

1. Construct a model volcano as follows. (You may choose to have students help you build the volcano, or simply introduce it to them as a surprise once it's completed.)
2. Use hot glue to attach the soda bottle to the center of the cardboard square.
3. Build a mountain shape around the bottle with the newspaper. Balling up the newspaper sheets and placing them near the bottle works well. Leave the top of the bottle open.

4. Cover the mountain shape with a layer of foil. Spray the mountain with flat black paint.
5. Now spray the mountain with spray adhesive. Sprinkle dirt over the mountain to give a realistic look.
6. Mix together 4 teaspoons (16 g) of baking soda, 8 tablespoons (120 mL) of water, 15 drops of red food coloring, and 4 tablespoons (60 mL) of dishwashing liquid. Pour the mixture into the top of the soda bottle.
7. Show students the overhead transparency of the *Three Types of Volcanoes.* Review how each type is formed.
8. Gather students around the volcano model. Ask students to tell you what comes out of a volcano (magma). Ask where the magma comes from (the mantle).
9. Tell students that you are going to make the volcano "erupt." Pour 5 tablespoons (75 mL) of vinegar into the soda bottle. Watch the foam "lava" pour from the mountain.
10. Have students describe how the lava would add new rock material to the mountain.

Follow-Up

Have students research the role of volcanic eruptions in making land fertile for agriculture.

Name ______________________________

Build a Volcano

1. What are the three basic types of volcanoes?

____

____

____

Procedure and Observations

2. Watch as your teacher makes the volcano erupt.

3. Diagram how the lava flowed out of the volcano.

Top View	Side View

Conclusion

4. How do volcanic eruptions affect Earth's surface?

____

____

____

____

Investigation 2

Crash Course Mountains

Materials

- student record sheet on page 45, reproduced for each student
- waxed paper
- modeling clay

Steps to Follow

1. Remind students that the Earth's crust is divided into plates and that these plates move around on the molten mantle.
2. Have students predict what happens when two plates crash together.
3. Give each pair of students two 6-inch (15 cm) squares of waxed paper and enough modeling clay to cover each square 1-inch (2.5 cm) thick.
4. Have the students flatten the clay on top of each waxed paper square about 1-inch (2.5 cm) thick.
5. Instruct students to set one square next to the other and push them together.

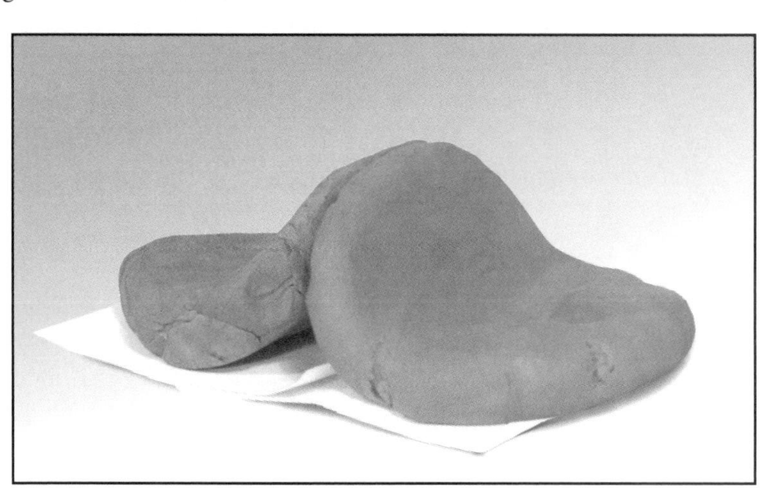

6. Have the students record their observations on their record sheets.
7. Explain that students have modeled how **folded mountains** are formed. When the two plates collide, they buckle and the crust pushes upward to form mountains. This is how the Himalaya Mountains (where Mt. Everest is located) were formed.

Follow-Up

Explain that the students have modeled one way that mountains can form. Have them research other actions that can create mountains (volcanic eruptions and faulting).

Name ____________________

Crash Course Mountains

Concept 4
Planet Earth
Investigation 2

Prediction

1. What do you think happens when two crustal plates collide?

____

____

Procedure and Observations

2. Flatten out your clay on the piece of waxed paper until it is evenly about 2.5 cm thick.

3. Push the two blocks together until they collide. Continue pushing them together until something happens.

4. What happened when the two slabs of clay crashed together? Draw and describe your observations.

____

____

____

Conclusion

5. What do you think happens when two crustal plates crash together? Support your answer with evidence from your experiment.

____

____

____

Investigation 3

Deposition

Materials

See advance preparation on page 40.

- student record sheet on page 47, reproduced for each student
- cloudy clay water
- shallow, clear plastic dishes
- magnifiers

Steps to Follow

1. Discuss the fact that water in rivers and streams carries particles of soil and rock suspended in it. Ask students to predict what happens to the particles when the water slows down or stops moving.
2. Divide the class into small groups. Give each group a shallow, clear plastic dish. Walk around the room, filling each dish about half full with cloudy water. Ask students to record their predictions on their record sheets.
3. Have students place their dishes aside over night.
4. The next day, have students observe their dishes and record what they see on their record sheets. (The clay particles should have settled out of the water.)

5. Lead a class discussion on how rock and soil particles are carried by water from one place and deposited in another place. If enough of these particles build up, land can be formed where before there was only water.
6. Challenge students to explain how soil carried in rivers and streams could settle out and build up land in certain areas.

Follow-Up

Have students research deltas. (As rivers dump their deposits into oceans, new land is formed along the coastline.)

Name ____________________

Concept 4 Planet Earth

Investigation 3

Deposition

Prediction

1. What do you think happens to sediment in streams and rivers when the water slows down or stops flowing altogether?

____

____

Procedure and Observations

2. Watch as your teacher pours water with sediment into your dish. Draw and describe what your dish looks like.

____

____

____

____

3. The next day, observe your dish again. Draw and describe your observations.

____

____

____

____

Conclusion

4. How can deposition affect Earth's surface?

____

____

Destructive forces affect landforms.

Prepare in Advance

Investigation 1: Collect several samples each of sandstone and limestone.

Investigation 3: Prepare enough "sandy" ice cubes for the class. Each group will need one cube. Prepare the cubes the day before the investigation by adding several pinches of sand to an ice-cube tray full of water and placing it in the freezer.

Teacher Information

Even as constructive forces are creating new landforms on Earth, destructive forces are wearing them down. Weathering and erosion are two forces that are constantly acting to reshape the land.

Weathering is the breakdown of existing rocks. **Mechanical weathering** involves the breakdown of rock by physical means including pounding, grinding, and cracking. Agents of mechanical weathering include water, wind, gravity—even plants and animals. As water freezes, it expands. Because water expands as it freezes, rainwater that collects in the cracks of rocks can break apart the rocks when it freezes.

Chemical weathering involves the breakdown of rock by chemical reaction. Chemical weathering turns the rock from one substance into another. For example, limestone reacts with vinegar (a weak acid) to release carbon dioxide gas, which can be seen as bubbles in the liquid vinegar.

Erosion is the displacement of soil and rock fragments from the place where they were created. Most commonly, this displacement is caused by the movement of water or glaciers.

As water runs from areas of higher elevation to areas of lower elevation (due to gravity), it picks up pieces of soil and rock fragments from the ground and transports them with it. In this way, Earth's materials are moved from one place to another. Where the fragments are deposited by the water, new land is formed. (See Concept 4.)

Glaciers are large bodies of ice that form as snow falls and piles up over many years. Gravity eventually begins to pull on these massive blocks, moving them downhill. As the glaciers move, they gouge out the land beneath them, carrying along large amounts of soil with them on their journey. Glaciers that form in river valleys often carve the valleys into a wider "U" shape as they move through them.

Wind and gravity are other agents of erosion. As wind moves across exposed areas of soil, it can carry the soil away. Gravity pulls rocks, boulders, sand, and soil from mountaintops to level ground over time.

Freezing Water Cracks Rocks

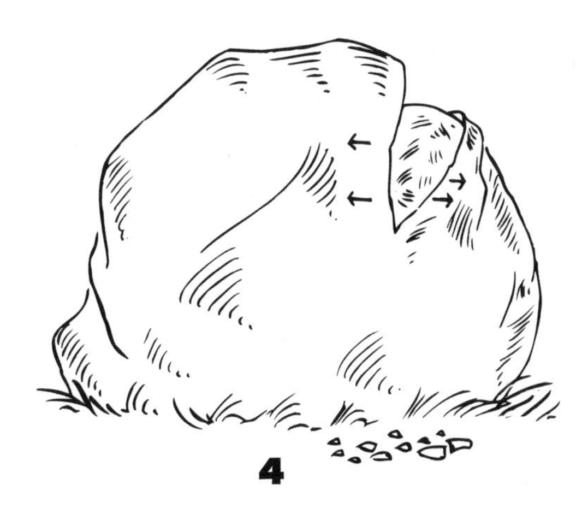

Investigation 1

Weathering Rocks

Materials

See advance preparation on page 48.

- student record sheet on page 51, reproduced for each student
- samples of sandstone
- samples of limestone
- dark paper
- magnifiers
- newspaper
- vinegar
- eyedroppers

Steps to Follow

1. Review with students the different ways in which rocks are formed. (See Concept 1.) Point out that rocks are also broken down by natural forces. Have students brainstorm a list of ways they think rocks can be broken down in nature. Record the list of ideas on the board.
2. Once the list is complete, divide the entries into two categories: **Mechanical Weathering** and **Chemical Weathering**. Discuss the differences between the categories.
3. Distribute the sandstone samples to each group of students. Have students rub the two samples of sandstone together over a sheet of dark paper.
4. Invite students to describe what it feels and sounds like as the rocks are rubbed together. Have students use a magnifier to examine the pieces of rock that collect on the paper. Tell them to record their observations on their record sheets.
5. Now distribute the limestone samples, a small amount of vinegar, the eyedroppers, and several sheets of newspaper to each group. Have groups cover their work areas with newspaper. Explain that vinegar is a weak acid.
6. Using the dropper, have students place several drops of vinegar onto the pieces of limestone.
7. Encourage students to use their magnifiers to observe any reaction that might be taking place. Have them record their observations on their record sheets.
8. Return to the list of weathering agents on the board. Challenge students to identify each of the processes they just modeled as mechanical or chemical weathering. (Grinding the sandstone modeled mechanical weathering. Placing vinegar on the limestone modeled chemical weathering.)

Follow-Up

To see the long-term effects of acid on rock, break off small pieces of the limestone and place them in enough vinegar to cover them. Have students examine the rocks over the next several days and describe any change. Afterward, invite students to research the causes and effects of acid rain.

Name ______________________________

Weathering Rocks

Procedure and Observations

1. Rub two pieces of sandstone together over a piece of dark paper. What did you see on the paper? Record your observations.

____

____

____

____

____

2. Place several drops of vinegar on the piece of limestone. What reaction (if any) did you observe after adding the vinegar? Record your observations.

____

____

____

____

____

Conclusion

3. How do mechanical and chemical weathering affect rocks?

____

____

____

____

____

Investigation 2

Water Damage

Materials

- student record sheet on page 53, reproduced for each student
- overhead transparency of *Freezing Water Cracks Rocks* on page 49
- plastic soda bottles, with caps
- water
- large resealable bags

Steps to Follow

1. Discuss with students what happens when rain falls. Where does it go? (It is soaked up by the soil, it collects on leaves, and it enters cracks in rocks.)
2. Remind students that when temperatures drop at night or during the winter, water freezes. Tell students they will do an experiment to see what happens to water as it freezes.
3. Divide the class into small groups. Distribute a plastic soda bottle to each group. Have the groups fill the bottle full of water and place the cap on securely. Tell them to make sure there is little or no air in the bottle. Then have them place the bottle in a large resealable bag.
4. If you live in an area where the temperature drops below 32°F (0°C) at night, have groups place their bottles outside overnight. If the temperature is not low enough, place the bottles in a freezer overnight.
5. The following day, remove the bottles from the freezer. Take them out of the bags. Have students examine them and record their observations on their record sheets.
6. Discuss the fact that no water was added to the bottles, so the water that was in there must have expanded and broken the bottles. Lead students to conclude that water expands (takes up more space) as it freezes.
7. Ask students what they think happens as water that gets into cracks in rocks freezes. Show students the *Freezing Water Cracks Rocks* transparency. Walk them through this aspect of mechanical weathering.
8. Have students complete their record sheets.

Follow-Up

Have students use what they learned in this investigation to explain why ice floats in a glass of water. (The ice is less dense than the water. That is, a given amount of ice weighs less than an equal amount of water. So it floats on the water, since less-dense materials float on top of more-dense materials.)

Name ____________________

Water Damage

Concept 5
Planet Earth
Investigation 2

Procedure and Observations

1. Fill a plastic soda bottle with water to the very top. Screw the cap on tightly. Put the bottle in a large resealable bag.
2. Place the bottle outside if it is cold enough, or in the freezer, overnight.
3. Observe the bottle the next day. What happened to the bottle? Record your observations.

____
____
____
____
____

Conclusion

4. Describe how freezing water can act as an agent of mechanical weathering in nature.

____
____
____
____
____
____

Investigation 3

Glacial Grind

Materials

See advance preparation on page 48.

- student record sheet on page 55, reproduced for each student
- "sandy" ice cubes
- aluminum foil

Steps to Follow

1. Hold a class discussion on how glaciers form and how the force of gravity pulls them slowly downhill. Tell students that in this investigation they will look at what effect glacial movement has on landforms.
2. Divide students into small groups. Distribute a sheet of aluminum foil and one sandy ice cube to each group. Explain that the sandy ice cube is a model of a glacier that has rock pieces embedded in it.
3. Have the students smooth the aluminum foil across their desks. (If the surface is a bit wrinkled, that's OK.)
4. Have students mimic the movement of a glacier across land by rubbing the ice cube across the surface of the foil. Have them record any observations on their record sheets.
5. Lead students to conclude that as glaciers move across the land, they erode away the earth beneath them. Explain that this material is carried away and deposited in another place at a later time. The process of moving earth materials from one spot to another is called **erosion**.

Follow-Up

Students can model deposition by allowing their ice cubes to melt and observing the pile of sand that is left behind.

Have students research continental ice sheets and the icebergs that break off of them.

Name ______________________________

Concept 5 Planet Earth

Investigation 3

Glacial Grind

Procedure and Observations

1. Spread out a sheet of aluminum foil on top of your desk.

2. Rub the ice cube your teacher gives you across the sheet of aluminum foil. What happened when you rubbed the ice cube across the foil? Draw and describe your observations.

____

____

____

____

____

Conclusion

3. What effect does glacial movement have on landforms?

____

____

____

____

____

Investigation 4

Stream Table

Materials

- student record sheet on page 57, reproduced for each student
- stream table
- wet sand
- plastic cup
- water

Steps to Follow

1. Remind students of their discussion of what happens to rain after it falls to Earth. (See Concept 5, Investigation 2.) Lead them to conclude that water moves from areas of higher elevation to areas of lower elevation. Explain that in this investigation they will find out how moving water affects landforms.
2. Divide students into small groups if you have multiple stream tables. If you have only one, do the investigation as a class demonstration.
3. Fill the stream table with several cups of wet sand. Smooth it out so that it is thicker at one end and tapers off at the other end. A few inches on one end should not be covered with sand.
4. Using your finger, trace a narrow path from the top of the sand to the bottom. Create a meandering path that snakes from one side to the other.
5. Fill the plastic cup with water and slowly dribble water at the top of the narrow path. Have students take turns doing this. Remind them to pour the water very slowly.
6. Have students observe any patterns in the sand. (The curved path may get curvier. A delta may form at the base of the sand.) Students should record their observations on their record sheets.
7. Discuss with the class what they observed. Challenge students to identify the processes of erosion and deposition in their model.

Follow-Up

Have students research the formation of the Grand Canyon by water erosion.

Name ____________________

Stream Table

Procedure and Observations

1. Observe the stream table as water is dripped into the top.

2. What did the stream table look like after water dripped on it for several minutes? Draw and describe your observations.

____
____
____
____
____
____
____
____
____
____
____

Conclusion

3. What effect does water movement have on landforms?

____
____
____
____
____

6

Soil is made up of a variety of components.

Prepare in Advance

Investigation 1: Have students bring in soil samples from around their homes, or provide a bag of potting soil for the class. Collect settling tubes (long, narrow plastic tubes with caps) for the class. If you cannot get settling tubes, collect and clean out small plastic soda bottles with caps.

Investigation 2: Prepare three different soil types. One should be very sandy (mix one part sand with one part dug-up soil), one should have a high clay content (dig up hard, dry soil with little organic matter and add extra powdered clay as needed), and one should be "ideal" (potting soil out of a bag). Prepare a bucketful of each sample, as the soils will be used in Investigation 3 to grow plants. Label the sandy soil "A," the clay soil "B," and the ideal soil "C."

Investigation 3: Purchase fast-growing grass (or other plant) seeds. You might check into Fast-Plants® from Carolina Biological Supply Company (www.carolina.com).

Investigation 4: Purchase a few dozen earthworms from a biological supply house or a bait shop. Each group will need 3 or 4 worms.

Teacher Information

Soil is made up of weathered rocks and organic matter. Weathering agents break rocks into smaller pieces. These pieces are mixed with decaying plant and animal matter. Soils vary greatly depending on the type of rock particles and the amount of organic matter they contain.

The quality of soil, for the purposes of growing plants, is determined by two principal factors:

- The soil's ability to hold the appropriate amount of water. Water retention is determined by the size of the rock particles making up the soil. Too many large particles (like sand), and the soil will drain too quickly and dry out. Too many small particles (like clay), and the soil will retain too much water and prevent gases from circulating around the roots.
- The soil's nutrient content. The soil must provide nutrients for the plants. These nutrients are found in decayed plant and animal material found in the soil. Generally speaking, the more organic matter soil contains, the more fertile the soil is.

When mixed with water and allowed to settle out, soil can be divided into its component parts. Larger, heavier particles will tend to settle out first. Smaller, lighter particles will settle out last. Thus a settling tube should show layers of heavier rock material toward the bottom and layers of lighter organic material toward the top. Settling tubes allow students to compare the makeup of different soil types by showing them the relative amounts of each soil component.

Earthworms improve soil by mixing it up and digging tunnels that allow water and gases to circulate freely around the plant roots. They also digest organic matter, breaking it down into a form more readily used by plants.

Subject: Evaluation of soil samples from newly purchased property

Dear Science Student,

Grow-A-Lot would like to provide the best plants to people who love to garden. In order to supply strong and healthy plants, we must find high quality soil. Our employees know a lot about plants, but they lack knowledge about soils. We have heard about the exceptional work your class has done in the field of soil analysis, and we hope that you will be able to help us.

We recently purchased an area of land in northwestern Washington, and have found large amounts of three different soil types. We would like to use these soils to grow our plants, but we don't know a lot about the soils. We hope you can help us create profiles of each sample.

Enclosed you will find samples of each soil type. Please conduct your experiments and return to us a report describing each soil sample, including a breakdown of its major components. We'd also like your opinion regarding which soil is best for growing plants.

Your report is vital to the success of our company. Thank you in advance for your help.

Sincerely,

Les Humus

Les Humus

President

Grow-A-Lot Plant Company

Investigation 1

Looking at Soil

Materials

See advance preparation on page 58.

- student record sheet on page 61, reproduced for each student
- soil samples (from students' homes or potting soil)
- newspaper
- magnifiers
- settling tubes or plastic soda bottles
- water

Steps to Follow

1. Have students cover their work areas with newspaper.
2. Distribute about ½ cup (120 mL) of soil to each student by dumping it in the center of the newspaper.
3. Have students use their senses to observe the soil. Magnifiers will allow them to look more closely at individual particles. Have students record their observations on their record sheets.
4. Distribute a settling tube or soda bottle to each student or group of students. Have them fill the tube ⅓ to ½ of the way with soil, and then add water up to 1 inch (2.5 cm) from the top. Instruct them to shake the tubes vigorously for several seconds.
5. Have students place their tubes on their desks and watch them for the next several minutes. (You may have to allow up to an hour for all the particles to settle out, depending on your soil samples.)
6. Have students draw and describe their observations on their record sheets.
7. Lead students to conclude that soil is made up of more than one kind of material.
8. Explain that soil is made up of rock particles and organic matter (bits of decaying plant and animal material). Challenge students to identify which layers in their tubes contain which types of material.

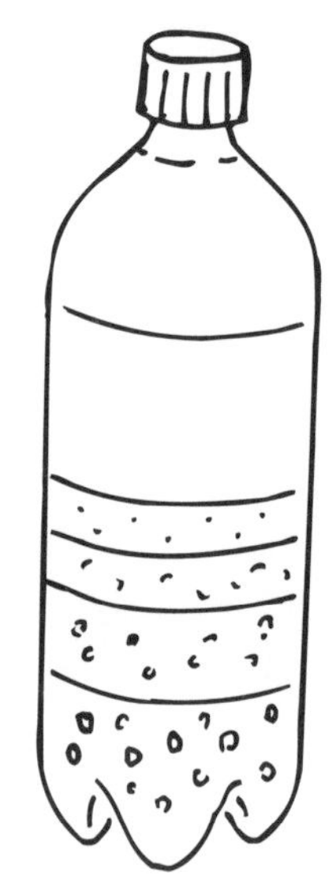

Follow-Up

Have students collect soil samples from other areas and create settling tubes of the samples. Do they notice any difference in the rock particles present? In the organic matter?

Name ____________________

Looking at Soil

Concept 6

Planet Earth

Investigation 1

Procedure and Observations

1. Cover your work area with newspaper.
2. Use your senses to observe the soil sample your teacher gives you. What is your soil like? Record your observations.

   ____

   ____

   ____

3. Fill your settling tubes according to your teacher's instructions. Shake them and then allow them to sit for several minutes.
4. Once the particles of soil have settled out of the water, observe the tubes again. Draw and describe what you see in your settling tube.

   ____

   ____

   ____

   ____

   ____

Conclusion

5. What did you learn about the makeup of soil in this investigation?

   ____

   ____

   ____

   ____

Investigation 2

Grow-A-Lot Requests Help

Materials

See advance preparation on page 58.

- student record sheets on pages 63–65, reproduced for each student
- copy of the *Grow-A-Lot Plant Company* letter on page 59, for each student
- samples of sandy, clay, and ideal soil (about ½ cup of each for each group)
- magnifiers
- settling tubes or plastic soda bottles
- water
- newspaper

Steps to Follow

1. Distribute copies of the *Grow-A-Lot Plant Company* letter to students.
2. Have students read the letter silently. Then discuss what the letter is asking students to do. Explain that today they will try to identify the components of each soil sample. In the next investigation, they will determine which soil type is best for growing plants.
3. Divide students into groups of two. Have groups cover their work areas with newspaper.
4. Pass out soil samples to each group, or invite groups to collect the samples from the front of the room.
5. Encourage students to design their own tests to evaluate the contents of the soils. Point out the magnifiers and settling tubes at the front of the room and invite students to use them as desired. (Settling tubes will allow students to see the proportions of each soil component in the samples.)
6. Once students have completed their analyses, have them compare their results with the results of other groups.
7. Lead a final class discussion of the soil sample components. Students should agree that sample A was sandy, sample B was predominantly clay, and sample C had a lot of organic matter.

Follow-Up

Have students start composing a class report to turn in to the Grow-A-Lot Plant Company. The report should include experimental procedure, raw data, and conclusions about the components of each soil sample. Students will be able to finish the report after Investigation 3.

Name ______________________________

Grow-A-Lot Requests Help

Investigation 2

Procedure and Observations

1. On the following pages, record the tests that you did on each soil sample and your data. Use diagrams as well as words to explain your tests.

Soil Sample A

Tests Performed:

Data:

Conclusion:

Soil Sample B

Tests Performed:

Data:

Conclusion:

Soil Sample C

Tests Performed:

Data:

Conclusion:

Investigation 3

Supporting Plant Life

Materials

See advance preparation on page 58.

- student record sheet on page 67, reproduced for each student
- soil sample reports from Investigation 2
- soil samples from Investigation 2
- plastic pots with drainage holes
- masking tape
- felt-tip pens
- grass seeds (preferably Fast-Plants®)
- water

Steps to Follow

1. Remind students of the letter they received from the Grow-A-Lot Plant Company, and of the data they collected in the last investigation. Pass out students' reports, or have them take the reports out of their notebooks and review what they learned about each soil type.
2. Tell students that today they will conduct an experiment to find the answer to the other question asked by the Grow-A-Lot Plant Company: Which soil type is best for growing plants?
3. Divide the students into groups of two. Have students cover their work areas with newspaper before collecting their planting materials.
4. Each group should fill three pots about ¾ full of each soil type (A, B, and C). Have them use the masking tape and felt-tip pens to label each pot with the correct sample letter.
5. Students should then sprinkle some seeds in each pot (approximately the same amount in each pot), cover with about ¼ inch (0.6 cm) more of the correct soil type, and pat down gently. Have them water each pot until water runs out the base of the pots. (You may want to have students step outside to do this.) Have them place their pots in an area of the classroom that gets direct sunlight at least part of the day.
6. Tell students that they should water the pots every few days as needed.
7. Have students monitor the plant growth over several weeks if possible, recording data on grass height, color, etc., on their record sheets.
8. Once students have completed their observations, hold a class discussion of the results. (Most likely, students will find that the nutrient-rich sample C produced the best plant growth. Results may vary.)

Follow-Up

Have the students complete their class report to the Grow-A-Lot Plant Company. Encourage them to include photos or drawings in the report. Post the finished report on a central bulletin board in your classroom or school.

Name ______________________________

Supporting Plant Life

Procedure and Observations

1. Plant grass seeds in each soil sample according to your teacher's instructions.

2. Observe the plants over the next several weeks. Record your observations in the spaces below. Use the back of this sheet for drawings of the plants.

Sample A	Sample B	Sample C
Date: **Observations:**	**Date:** **Observations:**	**Date:** **Observations:**
Date: **Observations:**	**Date:** **Observations:**	**Date:** **Observations:**
Date: **Observations:**	**Date:** **Observations:**	**Date:** **Observations:**

Conclusion

3. Which soil sample was best for plant growth?

____

____

Investigation 4

Earthworms

Materials

See advance preparation on page 58.

- student record sheet on page 69, reproduced for each student
- potting soil
- sand
- large glass jars with holes in lids
- black construction paper
- tape
- earthworms

Steps to Follow

1. Brainstorm with students a list of things plants need to grow (water, sunlight, nutrients from the soil, and air). Lead students to realize that one reason sample C in the last investigation produced such healthy plants was that the soil was light and air could get in between the soil particles to reach the roots. Also, rich nutrients were evenly distributed throughout the soil.
2. Have students speculate as to the role of earthworms in creating healthy soil.
3. Divide the students into small groups. Distribute a jar, a piece of black construction paper, and several pieces of tape to each group.
4. Have students bring their jars to the bags of soil and sand. Instruct them to create alternating layers of sand and soil in their jars. (The layers of sand should be thinner than the layers of soil. The top layer should be soil.)

5. Distribute three or four earthworms to each group. Tell the students to place the earthworms on top of their soil layers and place the lids on the jars. Make sure the lids have plenty of air holes poked out.
6. Have students wrap the black construction paper around the outside of the jar and secure it with tape.
7. Instruct students to observe their jars for two weeks (periodically removing the construction paper to look). Have them record their observations on their record sheets.
8. When the worms have mixed the layers of soil and sand, hold a class discussion of student observations. Lead students to conclude that earthworms improve soil by mixing the components together so that nutrients are evenly distributed. Students should also be able to see tunnels left by the burrowing worms. Explain that these tunnels allow air to circulate around the plants' roots.

Follow-Up

Have students research earthworm farms. How are the worms raised? Who buys them?

Name ________________________________

Earthworms

Concept 6 Planet Earth

Investigation 4

Procedure and Observations

1. Prepare your jar of soil and sand layers as instructed by your teacher. Add your earthworms to the jar and observe over the next two weeks. In the spaces below, draw what the jar looks like on each of your observation days.

Date:	Date:	Date:	Date:
Date:	Date:	Date:	Date:

2. What happened to the layers of sand and soil as the days passed?

____

____

Conclusion

3. How do earthworms help soil to be healthy?

____

____

____

Water cycles through Earth and its atmosphere.

Prepare in Advance

Investigation 2: Make "salt water" by mixing about 1.2 oz. (35 g) of salt with about 4 cups (965 mL) of water. You may need to make two or three batches.

Investigation 4: Cut the tops off several empty, clean 2-liter soda bottles. Save the tops and caps.

Teacher Information

Water is found throughout the Earth in various states (solid, liquid, and gas). On Earth's surface, it can be found as water in lakes, oceans, streams, and ponds. It can be found as ice in glaciers, and as snow on mountaintops. Just above Earth's surface, it can be found as water vapor (water in its gaseous state). Within Earth's crust, it can be found as groundwater, ice (permafrost), or even water vapor (at geysers).

The **water cycle** describes the movement of water from Earth's surface, into its atmosphere, and back again. The cycle is continuous, with no real starting point or ending point. Evaporation, condensation, and precipitation are the three principal processes responsible for moving water through the water cycle.

Evaporation: Evaporation is the change in state of a liquid to a gas. Water on the surface of the Earth evaporates and enters the air as water vapor. The vapor stays in the air until condensation occurs.

Condensation: When water vapor cools, it condenses to form liquid droplets in the air. The change in state from a gas to a liquid is called condensation. Condensed water droplets in the atmosphere can collect to form clouds. When the droplets become heavy enough, precipitation occurs.

Precipitation: As condensed water droplets gain in size and weight, eventually they become too heavy to float, and they begin to fall to Earth. Depending on the temperature, precipitation can take the form of rain, snow, sleet, or hail.

Water also moves through plants and animals on Earth. Plants take in water through their roots and release it through their leaves in the process of **transpiration.** Animals drink water and excrete it through their wastes.

Water is essential for life. Plants and animals must have water to live. The same water that quenches the thirst of a tiger in Nepal today may have moved through the roots of a swamp plant during the time of the dinosaurs. Water is never "used up," but instead constantly cycles through Earth's system to be used again and again.

The Water Cycle

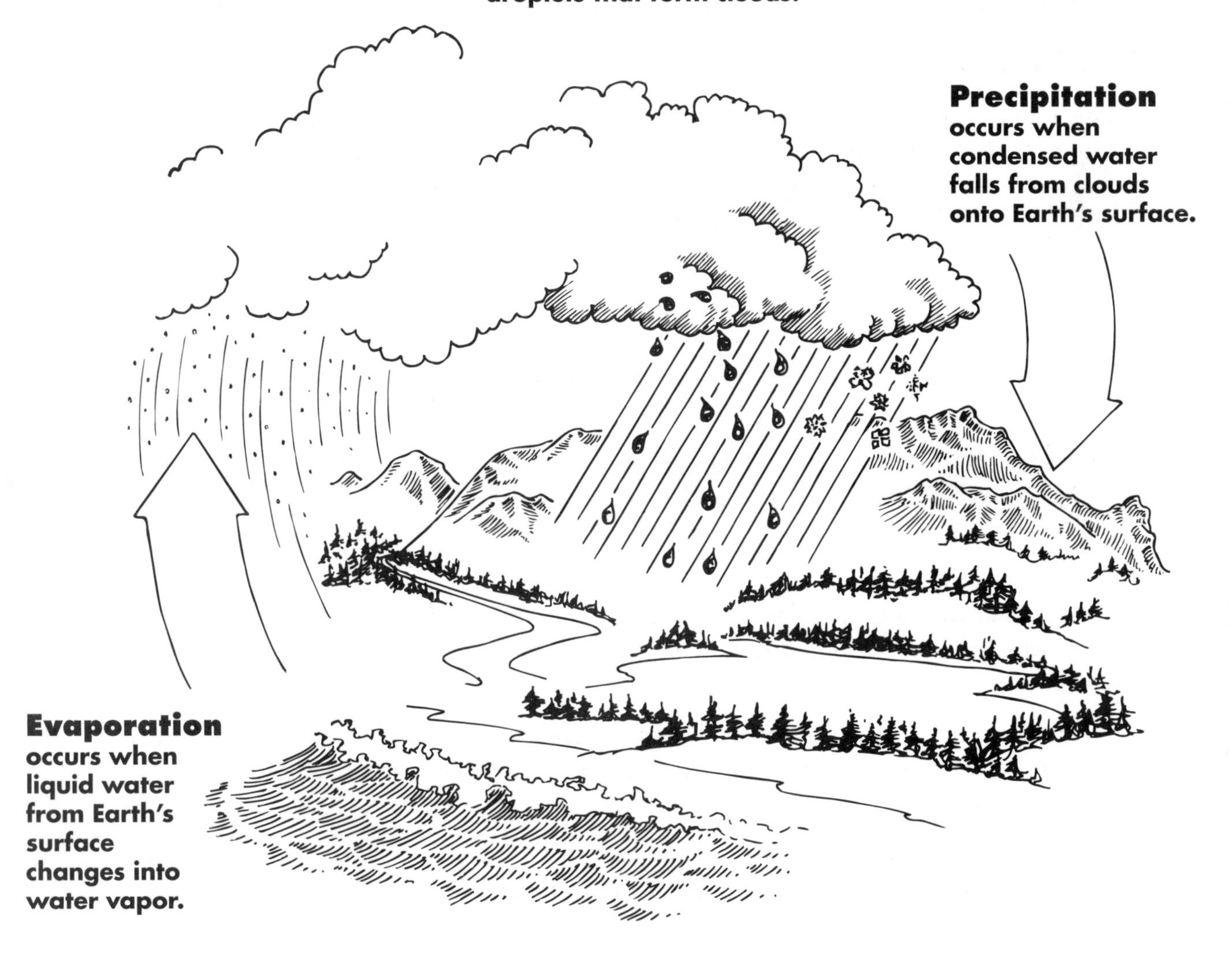

Investigation 1

Water, Water Everywhere

Materials

- student record sheet on page 73, reproduced for each student
- small slips of paper
- pencils

Steps to Follow

1. Tell students that today they are going to talk about water.
2. Divide students into small groups. Have groups brainstorm ideas about all the different places water can be found on Earth. Have them record each idea on a separate slip of paper.
3. Let them brainstorm and write for 15 minutes as you walk around and give hints to motivate their thinking.
4. Once activity has begun to die down, have each group place all of their slips into a pile in the middle of their table.
5. Instruct groups to separate their ideas into three groups: *solid* (ice), *liquid* (water), and *gas* (water vapor).
6. Discuss results as a class. You may want to create a class chart of all the different ideas discussed.
7. Now have the groups return all their slips to the middle of the table.
8. Have the students separate their slips into two new groups: *fresh water* and *salt water.*
9. Discuss student groupings as a class. Create a second class chart if desired.

Follow-Up

Have students bring their notebooks out into the school yard and write down every place they can think of where water can be found (surface water, air, plants, animals, soil, etc.).

Name ______________________________

Water, Water Everywhere

Procedure and Observations

1. With the other members of your group, brainstorm a list of places on Earth where water can be found. Write each idea on a slip of paper.

2. Place all the slips in the middle of your table. Then divide the slips into Solid, Liquid, and Gas groups, depending on what form of water each slip describes. Record your groupings on the chart below.

Solid	Liquid	Gas

3. Now divide the slips again according to whether or not the water source you described is salt water or fresh water. Record your groupings below.

Salt Water	Fresh Water

Conclusion

4. What did you learn about where water can be found on Earth?

____

____

____

Investigation 2

Disappearing Water

Materials

See advance preparation on page 70.

- student record sheet on page 75, reproduced for each student
- shallow plastic dishes
- paper towel
- water
- prepared salt water

Steps to Follow

1. Wet a paper towel and wipe it across the chalkboard so that it leaves a wet spot.
2. As the wet spot disappears, ask students where the water went.
3. Discuss with the class the process of **evaporation**. Bring forth the idea that water on Earth is constantly evaporating and entering the air as water vapor.
4. Ask students how they think the evaporation of salt water compares with the evaporation of fresh water. Tell them they will do an experiment to find out.
5. Divide students into small groups. Have groups place ½ cup (120 mL) of fresh water in a shallow dish, and ½ cup (120 mL) of salt water in a second dish. Instruct them to place the dishes in a spot that receives direct sunlight for part of the day.
6. Have groups observe the dishes every day for a few days, until all the water has evaporated.
7. Have students examine each dish and record their observations on their record sheets. (Both dishes should dry up at about the same time if they were placed in the same spot and both contained the same amount of water. Students should see a salt residue left behind in the saltwater dish.)
8. Discuss the salt residue in the saltwater dish. Challenge students to explain why it is there. (The salt in water does not evaporate, only the water does. All impurities are left behind in the dish.)

Follow-Up

Repeat the investigation using different container sizes in order to explore the relationship between surface area and evaporation rate. Use results to infer what happens in nature (evaporation rates in small-and-deep v. large-and-shallow bodies of water).

Name ____________________

Investigation 2

Disappearing Water

Procedure and Observations

1. Place an equal amount of salt water and fresh water in each of two shallow dishes. Set the dishes in a sunny spot, right next to each other.

2. Observe the dishes over the next few days, recording your observations on the chart below.

Date	Saltwater Dish	Freshwater Dish

3. What was left behind in the dishes once all the water had evaporated?

____

Conclusions

4. How did the evaporation rates compare in the two dishes?

____

____

____

5. What did you learn about evaporation in this investigation?

____

____

____

Investigation 3

Appearing Water

Materials

- student record sheet on page 77, reproduced for each student
- water
- food coloring
- ice
- drinking glasses
- paper towels

Steps to Follow

1. Remind students that water is constantly evaporating into the air. Ask them why it is, then, that the surface of the Earth isn't dry? Where does all the water vapor go? Tell students that today they will investigate the question of where the water goes.
2. Divide students into small groups. Give each group a dry drinking glass. Have them record their observations of the glass on their record sheets.
3. Have each group fill their glass ⅔ full of cold water. Place eight drops of food coloring into each group's glass. Use different colors across the class. Have students record the color of the water in their glass on their record sheets.
4. Now add several ice cubes to each glass. Have the groups observe their glasses for the next several minutes.
5. Call students' attention to the drops of water that begin to condense on the outside of their glasses. Ask them where they think the water came from. (Some may say that it leaked out of the glass.)

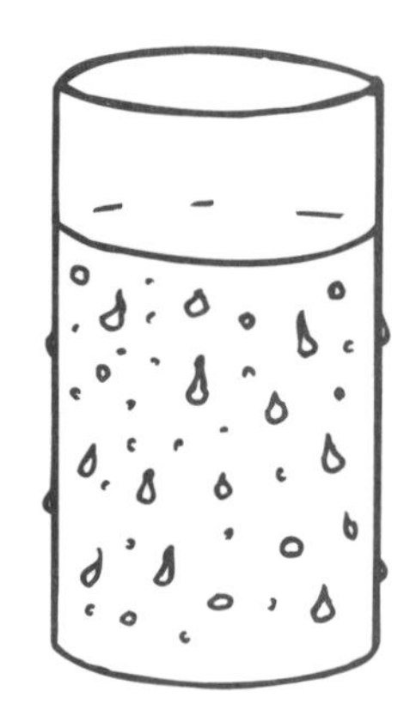

6. Have students use a paper towel to wipe some of the drops of water off the outside of the glass. Have them look at the wet towel and determine the color of the water on the outside of the glass (clear). You may want to have them stick another dry corner of the towel into the glass to confirm that the water in the glass is colored.
7. Challenge students to explain where the water on the outside of the glass came from. (It condensed from the water vapor in the air.) Point out to students that the ice lowered the temperature of the air around the glass, forcing the water vapor in the air to condense.

Follow-Up

A solar still uses evaporation and condensation to purify water. Have the students research solar stills. They may want to build one as a class.

Name ___________________________

Investigation 3

Appearing Water

Procedure and Observations

1. Observe the drinking glass your teacher gives you. Describe it. Is it wet?

____

2. Add water to your glass. Your teacher will then add food coloring. Describe what your glass looks and feels like now.

____

3. After your teacher adds ice to your glass, draw what it looks like below. Then draw what it looks like several minutes later.

glass just after ice is added

glass several minutes after ice is added

4. What color is the water in the glass?

____

5. What color is the water on the outside of the glass?

____

Conclusions

6. Where did the water on the outside of the glass come from?

____

7. What caused the waterdrops to form?

____

Investigation 4

Water Cycle Model

Materials

See advance preparation on page 70.

- student record sheet on page 79, reproduced for each student
- overhead transparency of *The Water Cycle* on page 71
- 2-liter soda bottles, with tops cut off
- soil
- small plastic cups
- water
- small terrarium plants
- spray bottle of water
- clear packing tape
- lamp

Steps to Follow

1. Review with students what they have learned so far about how water cycles through the Earth and its atmosphere.
2. Show students *The Water Cycle* transparency. Point out how they learned about evaporation and condensation in the last two activities. Go over the other parts of the diagram as a class.
3. Now tell students that they are going to build their own enclosed terrariums. A terrarium is a complete ecological system where water moves from one part of the system to another through the processes of evaporation, condensation, and precipitation.
4. Divide the class into small groups. Give each student one soda bottle (top and bottom). Invite them to fill the base of the bottles with soil and plants. A small plastic cup filled with water can model a lake in the mini ecosystem. Encourage students to use their imaginations.
5. Once the terrariums are complete, have students spray the insides with water and then place the tops on. Use masking tape to seal the two edges of the bottle together. Make sure the caps are on the bottles securely.
6. Place the lamp so that the light hits the terrariums.
7. Invite students to observe what happens inside the bottle over the next few days. (Students should see a humid environment, with waterdrops condensing on the sides of the bottle and running down to the soil.)
8. Have students draw a diagram of their terrariums on their record sheets. Have them label the water sources and processes of evaporation, condensation, and precipitation as they see them happening.
9. Challenge students to keep their terrariums going as long as possible. Periodically, take off the caps to allow gases to circulate. Also, spray water inside the bottle if the soil or plants appear to be drying out.

Follow-Up

Have students research weather patterns in different parts of the world. How do the weather patterns around the equator differ from those around the poles? How are these differences related to the water cycle?

Name ______________________________

Water Cycle Model

Concept 7 Planet Earth

Investigation 4

Procedure and Observations

1. Follow your teacher's instructions to set up a terrarium with your group.

2. Draw what your terrarium looks like below.

3. After a few days of observations, add labels to your drawing that show how water is evaporating, condensing, and precipitating in your terrarium.

Conclusion

4. How does your terrarium model what goes on in the natural world as far as the movement of water?

____

____

____

____